Internet of Things

Digital Tools and Uses Set

coordinated by
Imad Saleh

Volume 4

Internet of Things

Evolutions and Innovations

Edited by

Nasreddine Bouhaï
Imad Saleh

WILEY

First published 2017 in Great Britain and the United States by ISTE Ltd and John Wiley & Sons, Inc.

ISTE Ltd
27-37 St George's Road
London SW19 4EU
UK

www.iste.co.uk

John Wiley & Sons, Inc.
111 River Street
Hoboken, NJ 07030
USA

www.wiley.com

Library of Congress Control Number: 2017951079

British Library Cataloguing-in-Publication Data
A CIP record for this book is available from the British Library
ISBN 978-1-78630-151-2

Contents

Chapter 3. Introduction to the Technologies of the Ecosystem of the Internet of Things

Chapter 4. Toward a Methodology of IoT-a: Embedded Agents for the Internet of Things

Chapter 5. The Visualization of Information
of the Internet of Things   117

Adilson Luiz PINTO, Audilio GONZALES-AGUILAR, Moisés LIMA DUTRA,
Alexandre RIBAS SEMELER, Marta DENISCZWICZ and Carole CLOSEL

Chapter 6. The Quantified Self and Mobile Health Applications:
From Information and Communication Sciences
to Social Innovation by Design.   139

Marie-Julie CATOIR-BRISSON

Chapter 7. Tweets from Fukushima: Connected Sensors and Social Media for Dissemination after a Nuclear Accident 169

Antonin SEGAULT, Frederico TAJARIOL and Ioan ROXIN

Chapter 8. Connected Objects: Transparency Back in Play 189

Florent DI BARTOLO

Introduction

The development of connected and communicating objects has not stopped progressing as more and more objects are available in the market. This evolution of the Internet of Things (IoT) is creating more fields to be explored by the information and communication sciences, and renewing the risks of these new technological and digital changes in a "hyperconnected" world, via various connected objects (hyperobjects), which often have a dual capability: being connected and/or communicating while all the while carrying the expectation that they respond to user needs that are more and more demanding regarding services, communication and information.

The Internet of Things refers to these new objects/services, which are only a logical extension of the physical world into the digital world (hyperobject), and which generate a large amount of information, just as they receive it.

This work will present a collection of analyses, reflections and products/prototypes of connected/communicating objects (hyperobjects) as well as the prospect of studies and experimentation that these objects offer in the area of information and communication sciences. The data generated by these objects falls within the domain of Big Data, another related topic. Some texts are expanded and updated versions of texts from the International H2PTM Conference.

In the first chapter, the author Nasreddine Bouhaï defines the subject of the Internet of Things (IoT) and presents an overview of concrete examples of connected objects, whether they are intended for people's daily lives or for the world of art and culture. This non-exhaustive overview focuses on the massive influx of these new objects on the market. The question of intrusion into the private life of users is posed, as well as the question of security as a crucial point for the future of this ecosystem to come.

In Chapter 2, Ioan Roxin and Aymeric Bouchereau begin by presenting the historical and technological context of the evolution from the traditional web to the dynamic, social and semantic web and toward connected objects (CO). Secondly, they explain the definitions and concepts of the IoT based on examples of the IoT that are present in daily life.

In Chapter 3, Ioan Roxin and Aymeric Bouchereau focus more on the technological aspect of the IoT by presenting the elements related to context, architecture and protocols in the world of CO. They point out the major scientific problems to be resolved: the precise identification of each object in a network, standardization and finally, the normalization of data transfer protocols, machine-to-machine (M2M) communication, encryption and safety, the legal system and the architecture of the IoT.

The authors of Chapter 4, Florent Carlier and Valérie Renault, for their part, call on different paradigms of the IoT and the links that have been established in the literature between the IoT and multi-agent systems. In order to present a multi-embedded agent platform called Triskell3S, the authors demonstrate how the different paradigms and norms of the two areas can be respected and can coexist, in particular the MQTT protocol, the D-bus protocol and the FIPA-ACL specifications. Experimentation with this platform within a real context is done by an application of the IoT-a through a group of connected "screen-bricks" allowing the reconstruction of a wall of interactive and reconfigurable screens. We illustrate this application by revisiting the distributed eco-resolution N-Puzzle type (Taquin) algorithm and by taking it to the resolution of a Taquin video.

The visualization of information for the IoT is the subject of Chapter 5. The authors Adilson Luiz Pinto *et al.* return to the importance and the relevance of the use of visualization in the Internet of Things. The visualization and exploitation of the data coming from the IoT would increasingly interest users and companies. The integration of technology and the optimization of visualization of data is making it possible to display key information through graphics, tables, maps, etc. It has become possible to draw conclusions in a simple and visual manner, which is essential for businesses in order to be able to make decisions in real time, improve their performances, discover areas and anticipate problems so that they don't constitute a real risk for the company.

Chapter 6, by Marie-Julie Catoir-Brisson, focuses on the theme of the Quantified Self through the experience of Chris Dancy. The chapter is an analytical study for understanding what is involved in the integration of information technologies into people's everyday lives and how connected objects transform the relationship between the individual and his body and its representation and the human-machine relationship that this creates which accordingly increases the frequency of social interaction online. In order to grasp the multiple risks that this problem creates, an interdisciplinary approach is offered, an intersection of the analysis tools of semiotics, design and the anthropology of communication.

The authors of Chapter 7, entitled "Tweets from Fukushima: Connected Sensors and Social Media for Dissemination after a Nuclear Accident", Antonin Segault, Federico Tajariol and Ioan Roxin, are interested, through the study, in the dissemination of information via social media after a nuclear accident. This work is part of a research project on the use of social media in a post-nuclear accident situation, SCOPANUM (Strategies of Communication during the Post-Accident phase of a nuclear disaster through social Media). After having introduced the IoT (section 7.2) and recalling the elements of the role of social media in a crisis situation caused by a disaster (section 7.3), they describe the context, method and results of this study (sections 7.4 to 7.9).

In Chapter 8, Florent Di Bartolo examines modes of existence and operation in terms of the opacity and transparency of communicating objects. The author first tackles the sensitivity of connected objects to their associated environment and defines the type of relations that they establish with their users. He has then analyzed the illusion on which the Internet of Things is constructed: an illusion of transparency that presents communicating objects as enchanted objects and which artists and designers deconstruct to "open up" digital technologies and the data that they capture, disseminate and transform, to new forms of visibility.

In the ninth and final chapter of this work, Evelyne Lombardo and Christophe Guion reflect on the status of the body within the Internet of

Things. To do this, they begin by analyzing how the IoT transforms our relationship to the body in the context of e-health, then they pose the question of the traceability of the body through the integration of data. They then return to the concept of cloud data surrounding the body, to the interaction of this body within the network in order to study the body as a monitored body does not have the right to be forgotten. In the final section, they address the body as a communicating object between hyper-control and self-control.

The IoT: Intrusive or Indispensable Objects?

1.1. Introduction

Following Bill Gates' famous statement in the 1970s, "A computer on every desk and in every home," the world entered the era of computer science during the 1980s. This democratization became reality in developed countries, although not as much in third-world countries, which is a state of affairs identified by a digital and technological divide. New technological advances (computer science, telecommunications, miniaturization of electronics, etc.), led to the emergence of other solutions, new chips and electronic circuits, new computer systems and communication protocols, whose successful realization is the spread of mobile telephony and access to new compact and portable products. The smartphone is the prime example of this change; it now integrates all of the functions and services of a computer, making exchanges and communication accessible to a very large number of people. Moreover, with the connected watches that have appeared in the last few years, we are truly in the middle of the era of connected and portable devices.

Contrary to the development of computers and mobiles, whose concepts do not differ very much from one manufacturer to another (Apple, Windows, IBM, Dell, HP, etc.), the concept of the Internet of Things is broader and refers to a new way of living and managing current and professional affairs via the Internet. The environment is now more open for businesses and start-ups to innovate and offer new services and technologies. Nevertheless, the major players already have a head start in the area: like Cisco for networks, Google

Chapter written by Nasreddine BOUHAÏ.

for the management of big data, Microsoft for Cloud Computing, Intel for micro-processors, etc. It is clear that development and investment in the IoT, the businesses mentioned above, promising a future that is radiant but which remains nevertheless to be discovered and which will reveal whether this was a revolution or a passing technological fad. One of the goals of these objects is the transformation of uses or even creating new ones.

1.2. The age of miniaturization and technological progress

The development of computers and mobile telephony has been the technological duo of choice for several years. This has allowed the arrival in the markets of innovative projects, amazing and increasingly spectacular miniature-ization. The ENIAC[1] was the first electronic computer, occupying an area of a hundred square meters made to imitate a mechanical calculator[2]. An ultra-miniature version of the ENIAC computer, which is the size of a single integrated circuit chip, was developed by a research team from the University of Pennsylvania (Figure 1.1).

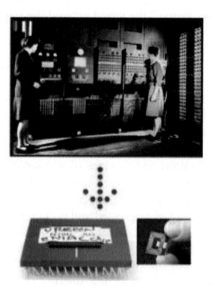

Figure 1.1. *The miniature version of the ENIAC*

---

1 Acronym for Electronic Numerical Integrator Analyser and Computer.
2 http://www.computerhistory.org/revolution/birth-of-the-computer/4/78.

The appearance of smartphones has been accompanied by enormous technological progress in the last decades, from the testing of the first mobile telephone, the Motorola DynaTAC 8000X[3] in 1973, to Samsung's most recent ultra-comprehensive and light smartphone[4], progress is exponential at different levels (Figure 1.2), computation power, design and ergonomics, energy consumption, etc. These advances have brought about a considerably profound change in the nature of the relationship humanity has with the objects and environment that surround it and a change to every person's everyday life and lifestyle.

Figure 1.2. *The evolution of mobile telephony*

1.3. The history of a digital ecosystem

The history of the Internet is enthralling and rich through its path of developing as an open system that is in perpetual motion. Despite its young age (it has been 25 years since the web was launched), the network has not stopped

---

3 actu-smartphones.com/24/le-premier-portable-au-monde-le-motorola-dynatac-8000x/.
4 www.samsung.com/fr/galaxys6/.

surprising us, thanks primarily to the work of communities of engineers and developers coming from different areas of study such as computer science, telecommunications and above all electronics. These are communities that connect to innovate and to respond to user needs in a collaborative and participatory spirit. Even if the origins and ideas of this network date back more than 50 years, a real enthusiasm was witnessed with the arrival of its best-known service, the web, which was put into operation back in the beginning of the 1990s. The revolution was provided by a multimedia navigation system with the development of the HTML language[5] that could integrate text, images and above all links between documents and fragments of documents. This extension of the Internet has taken on a new dimension, offering new experiences and new uses, as well as new difficulties, for navigation and tracking in a space of very dynamic and occasionally extensible links [BAL 96].

Since its conception, several layers have been added to the first version of the web. We can distinguish three essential steps in its development:

– the web 1.0: represented by the debut of the static and above all passive web of the 1990s, it offered basic navigation between pages of information whose purpose was documentary reference. This step was marked by the simplicity of the language used: HTML[6];

– the web 2.0, called the collaborative web, of the 2000s was the web of blogs, forums and CMS, with the web passing into active mode, with the users becoming actors and producers of content they played a contributing role and took forceful ownership of its new digital tools;

– the web 3.0: represents the current web of which semantics and connected objects are the two principle technologies.

From the web 1.0 to the web 3.0, to hypermedia [BAL 96] to the hyperobject[7], the Internet has gone from being based on information to being based on objects, from an Internet of links between documents, to one linking physical or digital objects (documents and information). It is a communicating and autonomous ecosystem, whose different objects are easily identified, and secure exchanges according to standardized protocols. These networks of objects[8] already pose the problem of traces of data

5 HyperText Markup Language, source https://www.w3.org/MarkUp/.

6 For Hypertext Markup Language, created by Tim Berners-Lee in the 1990s.

7 Principal theme of the International h2ptm Conference 2016, which was held in Paris, http://h2ptm.univ-paris8.fr.

8 The different connected objects are connected to small networks isolated from each other.

generated by the activities and exchanges of connected objects. Data to be exploited from the perspective of digital processing, according to approaches of knowledge engineering, another area concerned with the large masses of data otherwise known as Big Data.

1.4. Internet of Things, which definition?

The term Internet of Things (IoT) refers to a network that is more and more spread out, one of material objects connected to the Internet, identified and recognized, like all other traditional devices that we use every day, such as computers, tablets, smartphones, etc. Perceived these days as a new technological revolution, the Internet of Things is defined, according to Weil and Souissi [WEI 10], simply as:

> "The extension of the current Internet to all objects able to communicate, directly or indirectly, with electronic devices that are themselves connected to the Internet."

An official definition of the IoT remains to be found, a job for the actors in the domain, even if the overall concept and its components are well-known, such as the communication of data streams and associated protocols which remain a large open workshop.

Recently, tech giant Google has developed "Brillo", a platform for peripheral devices which handle the Internet of Things. It will be able to work with a very large optimization of the memory and processor, Wi-Fi and Bluetooth, it is derived from the "Android" operating system. Other companies have invested in the area, with Samsung's Artik, the Agile IoT platform from the manufacturer Huawei, intended for the IoT. Microsoft is not excluded, with a new version of its Windows 10. This shows the interest that large technology companies have in this new extension of the Internet.

1.5. The security of connected objects: the risks and the challenges

Data security is a crucial point and one of the greatest obstacles to the development of the IoT on a large scale. As with the Internet, security is a workshop in perpetual evolution, the problem is posed and is transposed logically onto the protection of data sent and/or received by a connected object and becomes a great technological challenge for the different actors in this new ecosystem.

We regularly see that digital insecurity is a recurring question, especially on the Internet network, affecting the hacking of websites, message servers, e-mail accounts and this is often done with a remote takeover of machines. This insecurity logically extends to the IoT. Like a connected computer, any connected object could be subject to hacking, a takeover, the installation of spyware, etc. With the impossibility of controlling and limiting the development of this ecosystem, it is necessary to look for and suggest security strategies for protecting the networks of these objects and to fill in the gaps in security detected.

The role of the telecommunications sector was and remains primordial for safeguarding the communication of these objects (object-object or object-person), as for the Internet, it is their responsibility to make as big an effort as possible to put in place solutions in the areas of security. A role that is just as important as that of software developers.

1.6. Protocols, standards and compatibility: toward a technological convergence

In this emerging market, a long-awaited consensus between the industrial actors in the domain is yet to arrive. It would make many products compatible with each other for the purposes of communication and the exchange of data. Currently each business uses its own technological solutions, a product manufactured by Samsung cannot exchange with one from LG, such as the automatic display of information from a television of one brand to a television from another brand. Task forces from several manufacturers[9] have recently discussed standards for objects connected to the Internet, to allow devices to mutually understand each other and determine the requirements regarding connectivity and interoperability between multiple devices. The question of norms and standards is central in the case of a need for technological convergence:

---

9 Includes a group of large companies such as Microsoft, Samsung, Intel and around 50 other companies.

"Normalization (and/or standardization) are notions which have become unavoidable with cultural, industrial, economic and especially digital globalization" [FAB 13].

The notions of norms and standards are present in Europe, and in America under the same name. It is understood that a norm is a frame of reference published by an official international organization for standardization such as the ISO[10], ECS[11], AFNOR[12] or the IEEE[13]. A standard can be described as a group of recommendations advocated by a group of representatives and informed users that is widely disseminated and used. HTML (W3C) format[14] for the web is the prime example of this type of procedure.

1.6.1. *The origins of some norms and standards*

Because the world of the IoT is obscured by a multitude of protocols, it is difficult to make an exhaustive list of them. A significant number of diverse solutions are ready to be developed quickly once norms or standards are integrated into future projects on a large scale. There are still many hypotheses to be confirmed in this rapidly expanding market. Some solutions are already on the market and others are in the process of development and validation, with the goal of standing out with their effectiveness and how simple they are to implement, an important point for small businesses and start-ups joining the IoT market, looking for communications solutions at the lowest cost until an agreement at this level has been reached. The goal will be to show the interest and usefulness of their products and to create a place and a name within the booming market[15].

In terms of communication, wireless is the best adapted to connected, and often portable, products. WiFi[16] and its variants are technologies that are increasingly popular at the moment, for short-/medium-distance

---

10 International Organization for Standardization: http://www.iso.org.

11 The European Committee for Standardization: http://www.cen.eu.

12 French Standardization Association: http://www.afnor.org.

13 Institute of Electrical and Electronics Engineers: https://www.ieee.org.

14 World Wide Web Consortium: https://www.w3.org.

15 Study by MARKASS (an expert in digital markets) from 2015 for the year 2016, "Connected objects and valorization des données – Tendances clés 2016": http://www. markess.com/.

16 Contraction of Wireless Fidelity.

communication[17] indoors and, with Bluetooth, as a short-distance[18] communication technology. Numerous protocols[19] supplement these two technologies, or even compete with them. Some have advantages such as a reduction in energy consumption[20]:

– WiFi direct[21]: unlike WiFi, which makes it possible to connect objects via an access point (an Internet box, for example), WiFi direct provides direct connectivity between two objects;

– Bluetooth LE/Smart[22]: considered complementary in relation to Bluetooth, it has low energy consumption, reduced coverage and a lower output. It is a solution for some types of connected objects;

– the Bluetooth aptx: a means of communication intended for audio broadcast by transcoding flows at a rate higher than 350 Kbit/s. A codec is used for the compression and diffusion of sound where the transmitter and the receiver must be compatible;

– the ZigBee[23]: this solution[24] offers connectivity with low energy consumption that is easy to embed within various connectable products, with a low bit rate that goes up to 250 Kbit, and a short coverage of around 100 meters;

– Near Field Communication (NFC)[25]: a solution for proximity communication (for a distance of a few centimeters). This protocol has its advantages: a miniature chip and the possibility of securing exchanges via an embedded encryption. Numerous uses, contactless payment, etc.;

– the Z-Wave[26]: this wireless protocol solution makes it possible to link several devices, it goes both ways, sending and receiving data. Its use is

---

17 From several dozen meters inside to several kilometers outside.

18 A dozen meters.

19 http://www.wi6labs.com/2016/03/16/quelle-technologie-radio-pour-les-objects-connectes-premiere-partie/.

20 http://www.wi6labs.com/2016/03/16/quelle-technologie-radio-pour-les-objects-connectes-twoieme-partie/.

21 http://www.wi-fi.org/discover-wi-fi/wi-fi-direct.

22 With LE standing for Low Energy.

23 Known by the name *IEEE 802.15.4*.

24 http://www.zigbee.org/.

25 For near field communication, source http://nearfieldcommunication.org/.

26 Source http://www.z-wave.com/.

adequate for home automation, with a coverage of 30 meters inside, to 100 meters outside;

– the Thread: established by Samsung and Nest Labs, is a competitor of the technologies mentioned previously, and consumes very little energy. It is a solution for home automation connectivity, to link different objects and devices in a network and to Internet. An alliance of several partners including Silicon Labs and Google gives it significant weight in the creation of future norms and standards.

1.7. Humanity, intelligence and technologies

1.7.1. *Crowdfunding as an aid to innovation*

Securing funding for making an innovative project a reality, especially for a young business without a history of activity and the multiplication of ideas and projects in the era of globalization, is not an easy thing. With the arrival of the IoT, the enthusiasm for this type of financing is without precedent[27]. Crowdfunding is an original principle (and an innovative approach), a fashionable solution for launching innovative projects with strong technological potential and for raising funds without too many constraints. The start-up Looksee[28], for example, is working on the Eyecatcher project, a smart bracelet that combines design, fashion and technological innovation (Figures 1.3 and 1.4).

Figure 1.3. *The Eyecatcher bracelet, display of notifications and messages in real time*

27 The statistics have been changing since 2011, source http://www.leguideducrowdfunding. com/a-savoir-mode-d-emploi/les-chiffres-du-crowdfunding/.
28 Source site Looksee: http://www.lookseelabs.com.

Figure 1.4. *The Eyecatcher in fashion mode*

The project's originality and innovation have already attracted more than 400 participants on the participatory platform Kickstarter, who have supported the project by raising hundreds of thousands of dollars, even though the creators of the project asked for only two thirds of this amount. Innovation lies at the level of low energy consumption thanks to its *e-ink* (digital ink) screen. Communication is done via Bluetooth with a smartphone application and makes it possible to send photos, designs and above all be programmed to send notifications such as e-mail, scheduling, etc.

1.7.2. *Participatory environmental sensors and citizens*

The Green Watch Project is a pioneering project in the field of connected objects and the result of research and development between an academic institution and industrialists. This project, of which the Paragraphe laboratory was one of the key elements for its realization, can be summarized as a group of participatory citizen sensors to measure the levels of ozone and noise in an urban environment. This project is part of an effort in the participatory and experimental sciences to rethink the relationship of the individual with his or her environment.

The technological and experimental aspect of the Green Watch Project consists of using two sensors: one for ozone and the second for noise. Geographic localization, necessary for getting the user's coordinates, is done with a GPS chip. The data is communicated via a mobile terminal (a telephone) with a Bluetooth chip.

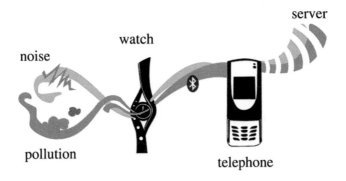

Figure 1.5. *Architecture of the Green Watch*

Figure 1.6. *Map of the data from Green Watch sensors*

This architecture[29] (Figure 1.5) provides the possibility of measuring, recording and communicating the data to an online processing and visualization mapping platform (Figure 1.6)[30].

29 http://www.linformaticien.com/actualites/id/6409/la-montre-verte-le-capteur-individuel-environ-nemental.aspx.
30 http://fing.org/?Le-succes-de-la-Montre-verte.

Connected objects for the environment have shown their effectiveness in many contexts, the automation of sampling in high-risk places, such as during nuclear disasters, as was the case in Fukushima. This was an example where citizens searched on the Internet and social networks in order to understand the dangers of the situation and act accordingly. Radiometers had been installed in the area of the accident to measure radiation in real time and were connected to the Internet network, and the results were published on social media [SEG 15].

1.7.3. *When digital art goes into connected mode*

Fictions d'Issy[31] is a generative and interactive novel developed at Paragraphe and presented during the Cube Festival, which was dedicated to digital creation, in 2005. The originality of the project's approach consisted of combining communication tools by connecting a text generator [BAL 06] to readers by means of a mobile telephone and displaying the texts generated on electronic information signs in the urban space of the town of Issy-les-Moulineaux (Figure 1.7), a first for this type of digital installation. This connected artistic work project was a pioneer in the field of living art[32]. The love story it tells is generated continuously by fragments of text of two characters who are evolving in the town's urban landscape.

Figure 1.7. *A display panel participating in the Fictions d'Issy installation[33]*

---

31 http://lecube.com/fr/fictions-d-issy-jean-pierre-balpe_444.

32 A form of expression specific to the digital medium: http://lecube.com/fr/ living-art-lab_154.

33 http://lecube.com/fr/fictions-d-issy-jean-pierre-balpe_444.

The principle of this connected work consists of the successive demands for the generation of fragments of texts (Figure 1.8) by readers via mobile phone calls using the keys on the keypad, with each one of the keys chosen influencing how the story unfolds and transforming the reader into an active participant in the story.

Figure 1.8. *Example of a fragment of the story displayed on an information panel in the town*[34]

1.7.4. *Home automation for a connected and communicating habitat*

In the past, the costs of constructing smart urban places were too high, and only accessible to a minority. The solution required the intervention of a specialized company, with a cumbersome process of integration and adaptation of devices. With the emergence of the IoT, home automation has made huge progress and has now become easy and inexpensive. You simply choose a central control device (*Home Hub*) compatible with a maximum of home automation objects[35]. Home automation is the field which has put the most items on the market, and it has not stopped developing since the first

---

34 http://lecube.com/fr/fictions-d-issy-jean-pierre-balpe_444.

35 Many products are available and compatibles such as those made by Philips, Netatmo, Bose, Osram, Sonos: http://www.usine-digitale.fr/article/smartthings-bras-arme-de-samsung-dans-l-Internet-des-objects-mise-sur-la-securite-de-la-maison.N347584.

days of the Internet, the evolution of networks and video surveillance. We now have a connected and communicating habitat, if not to say intelligent without exaggeration, because several areas[36] are involved in this revolution: security, energy, lighting, health, etc.

The smartphone has become the interface and the means of access and control for home automation. It is a simple and easy application interface available for the management of components. One pertinent example would be controlling entrance to and exit from the house, without the need for door keys, after the emergence of smart locks, such as the "Kevo" lock from the American company Kwikset[37] which makes it possible to order the opening or closing of a door remotely and without a key (Figure 1.9). The contribution of this type of object to everyday life is undeniable. If someone rings the doorbell, you are alerted via smartphone and it is no longer necessary to stay at home to let in a visitor (or a repairman, for example). Combined with a connected camera, it will be possible to hear and speak to him remotely.

Figure 1.9. *The Kevo smart lock from the American company Kwikset*

Smart cameras are part of the array of connected objects created in order to address the undeniable security needs of private individuals as well as professionals. They take up the torch of traditional video surveillance which consisted of setting up an IP camera and accessing it remotely. The new generation is clearly evolving: the HD Home from Withings[38] is a camera

36 A non-exhaustive list which is open to other domains.
37 www.kwikset.com/kevo/.
38 http://www.withings.com/eu/fr/products/home.

that integrates video recognition algorithms and night vision (Figure 1.10). Another function is audio analysis, which makes it possible to understand specific sounds such as the crying of children.

Figure 1.10. *The Withings HD Home intelligent video surveillance*

The interaction of these connected objects with the user allows object ↔ user communication via a smartphone, and an exchange of the data issued from the objects' environment by sensors, such as temperature, humidity, continuous measurement of the ambient air quality, and possibly allows other actions via communication with other objects.

1.7.5. *Connected objects, a step toward the enhanced human*

In Sweden, the use of local currency for payment has become almost obsolete. Donating to the church during mass or paying for a baguette or coffee is now done with ultra-modern methods, a smartphone in most cases, and, surprisingly, contactless payment by chips implanted under the skin of the hand (Figure 1.11). To make a payment, you just have to present your hand at payment terminal. Solutions are being tested by businesses for payments limited to cafeterias. A Canadian sports club now allows its members to be implanted with a chip with limited data to access the stadium[39], an experiment which says a lot about the concept of the augmented individual and which provides a glimpse

---

39 http://www.lapresse.ca/actualites/insolite/201604/26/01-4975217-une-puce-sous-la-peau-pour-entrer-au-stade.php.

of a future that is closer than we think. Other possibilities are being considered, since personal, financial and other data is loaded onto the electronic chip.

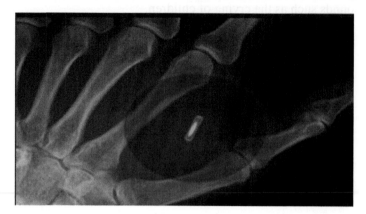

Figure 1.11. *Chip implanted under the skin of the hand[40]*

With the connected bracelets, activity sensors and the implant of a micro-chip for (auto-)surveillance, human beings are increasingly hyperconnected. The field of the "Quantified-Self" and the example of Chris Dancy[41] say a lot. This individu-Data [MER 13] shows the complexity that humanity could maintain with the masses of data that come from connected objects, and the way to interpret and use them in a healthy way.

This raises myriad questions about the connected human and questions about this way of life: security, private life, embedded personal data, etc. Social divisions are omnipresent when speaking about technological innovation, since older people who do not follow these technological changes very closely often find themselves on the fringes of these new uses and in difficulties socially, simply due to the fact that that societal evolution follows the majority and not the minority, which is the case in the digital society in which we have been living for some time now.

Another possibility for connected objects is to become a method of payment like any other. The company MasterCard, is working on ways to

---

40 http://www.journaldugeek.com/2015/02/12/societe-suedoise-implante-puces-rfid-sous-la-peau-de-salaries/.
41 http://www.chrisdancy.com/.

transform different fashion or other objects such as bracelets, rings, and smart watches[42] (Figure 1.12).

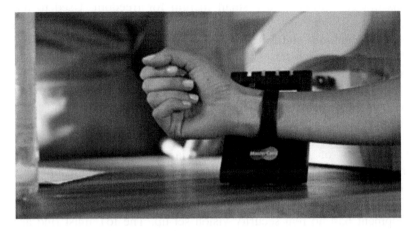

Figure 1.12. *Payment via a smart watch*

1.8. Conclusion

The technological revolution which has taken place over the course of recent years has totally changed the concept of computing. Previously, efforts were concentrated mostly on the development of office computers, smartphones, portable computers, tablets and other similar products. The change now operates on another level, with products that increasingly respond to consumer needs in daily life; products that are less bulky and with a refined design, with shapes that integrate logically into the user's environment, or physically in them or on them. This period was accompanied by advances in the area of the technology industry such as flexible touch screens, now used in smart watches, available on the market recently. Another dimension of use has been crossed, between computer products and these new miniature intelligent devices. It is no longer only a user-centered usage but one extended to his or her personal and professional environment. More and more objects make it possible to control the different elements of a habitat, a car, urban spaces, etc.

The rush toward the IoT bubble will continue, as with the first rush toward the Internet bubble, without worry about the medium- and long-term

42 http://www.stuffi.fr/mastercard-veut-transformer-nimporte-quel-objects-en-moyen-de-paiement/.

problems and risks that this provokes. The user's private life will be at the center of these preoccupations, increased control of the environment by various objects leaves a very small margin for freedom. Used to making choices himself, the user will see himself overtaken by some decisions whose consequences could be difficult to manage and correct. Hacking is another problem which raises a number of questions and leaves experts and users skeptical about the future. These products could easily be targeted by hacking with Ransomware[43] hacking, and put many people in extremely serious situations, whether through control of health devices or the control of private data.

The society of connected objects, via its experts, its users and its businesses, must reconsider and reflect on its future, to provide a perspective that is expert and reassuring at every level, especially security. This was the case with the Internet and its web service, which after several years of development had no real security vision set up. The IoT must find actors[44] from the same category as the Internet, to give users a display of confidence, and allow trouble-free development in this environment whose horizon remains to be examined with close attention by experts and researchers in different areas such as computer science, sociology, psychology, etc.

The users are the real actors and decision-makers in the market. Some see connected objects as a new fashion, a trend, which is just temporary for many of the objects already on the market or those to come, and which will fade away with time and combined with an abandoning of media interest in the subject, for others, it is really everyday objects which have a bright future in front of them. Between true innovations and illusory progress, the question will remain open for a long time as long as scientific and economic studies do not validate the economic, social and technological reality of this market.

1.9. Bibliography

[BAL 86] BALPE J.-P., *Initiation à la génération de textes en langue naturelle, Exemples de programmes en Basic*, Eyrolles, Paris, 1986.

[BAL 96] BALPE J.-P. *et al.*, *Techniques avancées pour l'hypertext*, Hermes, Paris, 1996.

---

43 A malicious software program that hacks into personal data to demand payment.
44 Such as the World Wide Web Consortium, Industrialo Internet Consortium.

[FAB 13] FABRE R., HUDRISIER H., PERRIAULT J., "Normes et standards: un programme de travail pour les SIC", *Revue française des sciences de l'information and de la communication,* no. 2/2013, available at: http://rfsic. revues.org/ 351, consulted June 15, 2016, made available online January 1, 2013.

[MER 13] MERZEAU L., "L'intelligence des traces", *Intellectica,* no. 59, pp. 115–135, 2013.

[SEG 15] SEGAULT A., TAJARIOL F., ROXIN I., "Tweets de Fukushima: Capteurs connectés et médias sociaux pour la diffusion de l'information après un accident radiologique", *Actes du colloque H2PTM'15,* ISTE Editions, London, 2015.

[WEI 10] WEIL M., SOUISSI M., "L'Internet des objets: concept ou réalité?", *Annales des Mines – Réalités industrielles,* no. 4, pp. 90–96, 2010.

The Ecosystem of the Internet of Things

2.1. Introduction

The computing world has experienced an exponential evolution over time, from the first mainframe computers to cloud computing, not to mention workstations and mobile computing. The processes involved in these transformations have ended up making computing and communication networks ubiquitous. Objects in the physical world communicate with the digital world (computing) by becoming connected objects (COs) with enhanced functions. COs and devices make it possible to store, transmit and process data taken from the physical world, touching on many aspects of human life: food, agriculture, industry, health, wellbeing, sports, apparel, habitat, energy, video surveillance, pets, etc. According to a study done by the Swiss Federal Institute of Technology, Zürich, in 10 years (2015–2025), 150 billion objects will be connected worldwide; the volume of data generated will double every 12 hours (versus around every 12 months in 2015). Technological innovation is plentiful and the market will then sort between gadgets and truly useful communicating objects.

Passive or active, identified and uniquely identifiable, COs have a direct or indirect link with the Internet. We are talking about the Internet of Things (IoT). The domain leads to major challenges with regards to our capacity to construct an optimal and safe ecosystem for the IoT. The "physical objects"/"associated virtual intelligence" pair, whether it is embedded, distributed or hosted in a Cloud, will lead us toward technologies or methods

---

Chapter written by Ioan ROXIN and Aymeric BOUCHEREAU.

of software design related to artificial intelligence and also toward the science of complexity.

In this chapter, we will first present the historical and technological context of the evolution from the traditional web to the social and semantic web and to COs. Secondly, we will clarify the definitions and concepts of the IoT by drawing on examples of its presence in our daily lives.

2.2. Context, convergences and definition

The emergence of the IoT fits into the logical progression of elements that have contributed to the construction of computing and of communication networks as they are known today: the birth of the World Wide Web, the democratization of the web, the passage from analog to digital and technology convergence.

2.2.1. The Internet Toaster or the first connected object in history

Several years after the first steps of the Internet and TCP/IP protocols, "Dan Lynch, President of the Interop Internet networking show, told John Romkey at the 1989 show that he would give him star billing the following year if he connected a toaster to the Internet" [STE 15]. In 1990, with the help of Simon Hackett, a friend, Romkay successfully connected a Sunbeam Deluxe Automatic Radiant Control Toaster. Controllable via the Internet and the TCP/IP protocols, it was possible to turn the toaster off and on remotely. The toasting of the bread depended on the operating duration of the device. One year later, the famous toaster was augmented with a robotic arm that could be controlled remotely and which could pick up a piece of bread and place it in the machine [STE 15]. Other manipulations of this kind were introduced in the following years, for example, the first use of a webcam, the famous *Trojan Room Coffee Pot* (1991) [KIE 01]. Another example is the soda machine at the Computer Science Department at Carnegie-Mellon University where, to avoid long trips to an empty machine, students added sensors that made it possible to find out the contents of the soda machine remotely[1] [CAR 98]. In addition to demonstrating the inventiveness and

---

1 On April 1, 1998, an RFC published by the IETF detailed the operations necessary to connect a coffee machine to the Internet (Source: http://www.rfc-editor.org/rfc/rfc2324.txt).

ingenuity of their creators, these experiments provided a foretaste of what we call in the 21st Century, the "Internet of Things". All of these experiments are very much in line with the ideas conveyed by this concept: connecting everyday objects to the Internet with the help of information and communication technologies, in order to enrich their functionalities. Thus, we talk about "connected objects". The birth of the Internet, its expansion and the pervasiveness of computing are some of the elements which led to the theorization of the IoT.

2.2.2. From the Internet of computers...

Derived from the concept of Internetting ("interconnecting networks") and based on the foundations laid by the ARPANET packet transfer network, the Internet is a network of networks. This innovative system was developed in the United States in the 1970s with the goal of unifying connection techniques and facilitating the sharing of resources between different computers and operating systems. However, the Internet would shift away from its original goal to connect different communication networks (for example ARPANET, communications satellites, radio communications).

Starting in the early 1970s, Vinton Cerf and Robert Khan developed the TCP (Transmission Control Protocol) and IP (Internet Protocol) protocols on which the Internet is built. The IP protocol is in charge of the delivery of the packets that it transmits to the recipient using an IP address. Thus, each terminal[2] connected to the network was assigned a unique address (IP address). As for the TCP, it is responsible for the reception of packets, guaranteeing the order and successful delivery of the packets transmitted from one host to another. It was in January 1983 that the TCP/IP pair of protocols was adopted and also with them the word "Internet" [HAU 03]. Following this event, the network was deployed on a larger scale[3] to finally become a worldwide network of interconnected computers.

A multitude of services were developed around the Internet: the transfer of files (File Transfer Protocol or FTP), electronic messaging (e-mail), delayed-time discussion forums (newsgroups), dialogues in real time (Internet Relay Chat) as well as the web or World Wide Web (WWW).

---

2 The terms "terminal" or "host" here refer to all devices connectable the Internet network (for example printer, server, office computer or router).

3 France first connected to the Internet network on July 28, 1988.

2.2.2.1. *Identification of resources*

On March 13, 1989, Tim Berners-Lee proposed a hypertext[4] system to facilitate the sharing of documents within the European Council for Nuclear Research (CERN)[5]. Convinced of the project's value, Robert Cailliau joined Tim Berners-Lee in 1990 and they both created the WWW[6]. The same year, the first web browser and editor christened the WorldWideWeb were introduced as well as the first web server, called "CERN httpd" [CON 00]. It made it possible for anyone to consult resources ("web pages") on distant sides with the help of a browser. Each web page corresponded to a node and it was possible to navigate from one page to another by clicking on a "hypertext link" or "hyperlink" (this type of link makes it possible to move around in a hypertext system).

This type of navigation between web pages was made possible using HTML language (HyperText Markup Language) to structure the information on web pages. In fact, HTML language was created by Tim Berners-Lee at the beginning of the 1990s to represent web pages. HTML is a computer markup language, that is to say, one that makes it possible to specify the structuring of information contained in a web page (for example shaping the text, creating formulas and tables, including images and videos) with the help of markups and hypertext links[7].

Users consult these web pages by means of a web browser or "HTTP client". The latter made it possible to "download" distant web pages via HTTP protocol (HyperText Transfer Protocol), invented by Tim Berners-Lee in 1990 as part of the development of the WWW. HTTP ensures client-server communications: a client communicates with a server using HTTP to transfer the resources requested by the client from the server. There is also HTTPS, a secure version of the protocol.

---

4 The term was invented by Ted Nelson in 1965 as part of project XANADU. The web was inspired by this project, which was about an information system imagined to instantly share information thanks to computers.

5 The same year, in 1989, CERN connected to the Internet network and recorded its first external connection.

6 In 1990, a widely-circulated joke said that the acronym WWW stood for "World Wide Wait" in reference to how slow the network was as a result of its wide popularity.

7 The language has gone through several versions over the years including HTML 4 (since 1997) which was used for many years until the release of HTML 5 in 2014.

Finally, to establish communication with an HTTP server and access its resources, the client must have the server's web address. Another invention of the founders of the web, these addresses are strings of characters that uniquely identify each web page. They generally take the following forms: http://www.example.com or www.example.com, and are based on hypertext links. In reality, they are URIs (Uniform Resource Identifier) or IRIs (International Resource Identifier), a standard which defined the syntax of addresses. A URI can be the "locator" type otherwise known as a URL (Uniform Resource Locator) or the "name" type known as a URN (Uniform Resource Name). The URL identifies a resource on a network by describing its location[8] while the URN allows the identification of a resource by its name without having to reference its location[9]. This naming system gives a unique identifier to all the resources present on the web; this way each one of them is recognizable and accessible.

2.2.2.2. Evolution of the web

Over time, the web has gone through several iterations. Thus, if yesterday's web was the traditional web (web 1.0), a web of documents and firmly static, today's web (web 2.0) is a social and dynamic web. The web of tomorrow (web 3.0) will be a semantic and collective web linked to the IoT. For the web of the day after tomorrow, researchers and futurists talk about a web that will integrate intelligent augmented reality (web 4.0) and the symbiotic web (web 5.0).

2.2.2.2.1. Web 1.0

The first version of the web (in the 1990s), also called the traditional web, was one of documents, with passive and static functioning much like that of a library where Internet users went to consult resources. Since the distribution of information was at the base of the web 1.0, users had only a passive role, a status of "spectator" and could not participate in any way in the creation of new resources.

2.2.2.2.2. Web 2.0

Since the year 2000, the web has been dynamic and collaborative, allowing Internet users to be true actors (the participative web). In addition to consulting resources online, which became more diverse (for example

---

8 http://example.com is a URL.

9 "urn:isbn:978-2-7637-8405-2" is a URN designating a book whose ISBN number is ISBN 978-2-7637-8405-2.

photos, texts, videos, music), web users can create content via blogs, wikis or social networks. The web became a space for socialization where Internet users communicated, shared and created links (for example on Facebook, LinkedIn, Snapchat or YouTube).

Another attraction of the web 2.0 was the possibility of developing and creating very specific applications from data distributed by certain platforms. Service providers such as Instagram, Flickr or Google offered developers the possibility of using raw data generated by their services via APIs (Application Programming Interface)[10]. A directory dedicated to API, constantly updated, is found at the address http://www.programmableweb.com/.

2.2.2.2.3. Web 3.0

The third generation of the web refers to the semantic web (or web of data) whose goal is to make the resources present on the web more easily usable and understandable ("readable") by machines. The idea is to gather the information in a useful way, as if in a gigantic database where everything is described in structured language. To do this, in addition to the shape and the structure, the semantics of the resources, namely the knowledge about these resources and the relationships between them must be described. If the RDF (Resource Description Framework) model is the *lingua franca* for sharing metadata, it is ontologies that are used to describe the semantic constraints and reasoning. In order to create, exchange, merge, expand and connect different ontologies, the W3C (World Wide Web Consortium)[11] proposes OWL (Web Ontology Language) as the ontology language for the web. With precise and complete ontologies, computers can act as if they "understand" the information that they transmit. The semantic web therefore has the goal of giving machines a means that allows them to implement computational shortcuts in order to simulate human reasoning.

The IoT accompanies this generation, since a closer link between the physical and virtual worlds was born when the mobility of Internet users increased through a multitude of connected devices (for example smartphones, tablets, smart watches or connected cars) in the web 3.0.

---

10 API: programming interface, "an interface containing the interface necessary functions for the development of applications" (Source: http://granddictionnaire.com).
11 The W3C is a standardization organization dedicated to web technologies.

In his thesis published in 2011 [TRI 11], Vlad Trifa develops the concept of the "Web of Things" (WoT), namely the integration of COs into the Internet network as well as the WWW. In the WoT, Trifa sees the alliance of the social, programmable, semantic, physical and real-time webs, many particular facets that the WoT would have.

2.2.2.2.4. Web 4.0

The web 4.0 is characterized by the emergence of a symbiosis between humans and machines as well as an intelligent entity [PAT 13]. Machines in the web 4.0 will be at least as performing as the human brain, capable of understanding the content present on the web (in keeping with the semantic web) and reacting in an appropriate way, taking into account the user's expectations. In other words, interactions between users and machines will be of higher quality, since the web and computer systems would have abilities better allowing them to grasp the reasoning behind content and user requests. For example, web platforms can personalize their interfaces according to each user's habits and can individualize the business-client dialogue. We can currently see the premises of this type of interface through the recommendations and suggestions of products offered by retail sites such as Amazon.

2.2.2.2.5. Web 5.0

Web 5.0, the symbiotic web, refers to the transformations appearing along with the densification of links, addresses [IP] and more generally of the Internet network. Arriving at a critical threshold, the concentration would be so strong that new properties would emerge, bringing us to a "transition"[12]. Futurists and trend analysts such as Joël de Rosnay and Ray Kurzweil imagine the next major evolution in computing and the Internet. Joël de Rosnay proposes several terms to define what the computing world of tomorrow will be: web 5.0, "Symbio-Net" or the "biotic" and "intelligent environment" [DER 10].

Joël de Rosnay sees a merging between biology and computing, a marriage of domains which he calls "biotic." From this merging, a symbiosis

---

12 This transition is referred to as the "technological singularity" and describes a certain point where exponential technological growth move to a higher stage and "all of our current predicative models will be null and void." Ray Kurzweil develops this notion in his book Humanité 2.0: la bible du changement (2007). In 2008, Peter Diamandis and Ray Kurzweil created Singularity University (http://singularityu.org/), whose slogan is "Making the impossible possible!".

between Man and the Internet would be born that would become so pervasive that: "[…] [Man] will no longer be on the Internet, but in the Internet […]." The biotic is principally characterized by the development of biosensors which make it possible for machines to receive information, from the human body.

Figure 2.1 illustrates the evolution of human-machine interaction from graphic interfaces (GUI – Graphical User Interface or WIMP – Window, Icons, Menus, Pointer), through tactile/acoustic interfaces to gestural/ intentional interfaces.

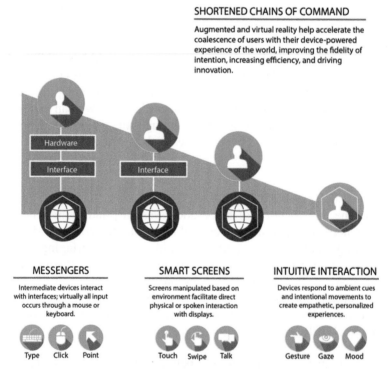

Figure 2.1. *Evolution of human-machine interaction (adapted from [KUN 16])*

The Symbio-Net, the fifth generation of the web, can be considered as a three-dimensional (3D) space where individuals, on their own, can navigate and consult resources [PAT 13]. This 3D world is modeled thanks to the processing power and memory of all interconnected devices, Smart Communicators (for example tablets, smartphones, personal robots), which generate virtual avatars of their users. In addition, the web 5.0 will be able to know emotions of the individuals traveling through the Symbio-Net and they can interact emotionally with the content that they consult [PAT 13]. It is an evolution where the border between the neuron and computer chips barely exist anymore. The biggest change concerns the messages sent and the path used, which would no longer just be from the brain to the machine, but also from the machine to the brain.

"Biology becomes technology, technology becomes biology, humans become more and more roboticized, robots more human and progressively supplanting us, driving us toward a new world, a new humanity". (Béatrice Jousset-Couturier, Le transhumanisme, Eyrolles, Paris, p. 188, 2016)

According to Ray Kurzweil [KUR 07], "the machines of the future will be human even if they aren't biological [...]. The majority of our civilization's intelligence will finally be non-biological. Our civilization will remain human, nevertheless, it will be, in many respects, much more exemplary than what we consider human today".

2.2.2.3. Convergence

The ideas that drive the IoT have existed since the 1990s and coincided with the beginning of the Internet. Despite some early experiments, the phenomenon was only emphasized after the year 2010 [MIC 15]. The development of the IoT happened alongside the manifestation of several types of convergence (digital, technological, services, networks, devices and policies). These convergences will push usage and technologies toward a change in paradigm concerning the Internet and computing in general.

The massive changeover from analog to digital stimulates the convergence of communication technologies and multiple sectors have been transformed (for example the audiovisual, television and telecommunication sectors).

2.2.2.4. Digital convergence

The "digital convergence" participated in the generalization of computing and of the WWW as well as its application to several domains such as medicine, administration or photography. It involves the fusion of several elements that previously functioned independently from each other: content, support and transportation.

Firstly, content refers to information that is legible and understandable to Man, that is to say, a series of bytes representing photos, video cassettes or paper documents. This information has been digitized: we have gone from analog to digital.

The support, the means used to read, listen to or watch a piece of content, was dependent on the type of content in the analog world. With digital, the support no longer exists in different types of memories and protocols for interpreting the content. A type of content no longer requires a particular support. As a result, a video can be viewed on a smartphone, a tablet or an office computer. Furthermore, on a tablet, for example, we can listen to music, watch a film, read a book or check our e-mail.

Finally, transportation evokes the capacity or the process used to carry the content to the user. Digitized and dematerialized, the content becomes transportable through any network and can be watched, read, manipulated or listened to from any place connected to the Internet. Geographical position is no longer a determining factor in the accessibility of content. Nor, moreover, is time. The term ATAWAD (Any Time, Any Where, Any Device) perfectly symbolizes the digital transformation of our society and expresses the possibility of getting information at any moment (anytime), in any location (anywhere), from any piece of equipment (any device) connected to the Internet.

The digital convergence is the fusion of devices designed for a single type of content (for example the camera, Hi-Fi, VCR, television) thanks to their digitization [WCI 12]. The convergence has allowed computers as well as other devices (for example smartphones, MP3 players, tablets and personal assistants) to be endowed with new functionalities such as playing videos or music. This phenomenon has especially allowed multimedia systems to develop. Another consequence consists of the disappearance of previously impermeable borders between sectors of activity. The digitization of content is becoming more and more systematic, the Internet network and computing are now established as part of telephony, photography[13], journalism, paper documents generally and networks (certain models of portable telephone, MP3 player, a portable speaker or even a lamp with WiFi receivers).

---

13 On January 13 and 20 and March 26, 2006 respectively, Nikon, Konica Minolta and Canon announced that they had abandoned the development of silver halide photography in favor of digital photography.

At a conference held at the Collège de France in 2015, Joseph Sifakis, Professor at the Federal Institute of Technology Lausanne (EPFL), addressed the applications of digital convergence. Figure 2.2 illustrates the four domains which have been changed by the switch to digital: services, networks, policies and devices.

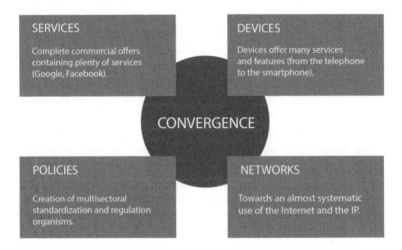

Figure 2.2. *Convergence of services, devices, networks and policy (adapted from Joseph Sifakis [SIF 15])*

Services should be interpreted as the multiplication of complete commercial offers containing a multitude of functions, for example those offered by the companies of the GAFA[114]. For example, Google doesn't just offer a search engine, nor does Facebook offer a social network, but rather offers uniting a large panel of applications. The offers are more and more spread out and touch on aspects that are sometimes very disparate[15], to the

---

14 The term GAFA is an acronym of the names of the four giants of the Internet (Google, Apple, Facebook, Amazon) also known as the Big Four. Often, we see the acronym GAFAM (M for Microsoft) or GAFAMA (A for Alibaba). After the GAFAMA, other major actors appear under the acronym NATU (the acronym dates to summer 2015) which groups together the four companies symbolic of digital "disruption": Netflix, Airbnb, Tesla and Uber.

15 The company Google, whose primary offering was the search engine, has greatly diversified (for example the search engine, autonomous car, biotechnology, health, home automation). The Alphabet Inc. conglomerate of businesses created in 2015 gathers all the services previously held by Google Inc. (for example Google, Google X, Google Capital, Nest, Calico).

point of creating multifunction platforms that finally mean users are calling on very few different companies.

Internet and TCP/IP protocols are used on a large scale in diverse applications and services and are now the default choice when it comes to establishing communication between several actors. For this reason, integrated into different communicating devices, IP protocols are becoming a universal means of communication, sometimes exploited beyond their limits[16]. This is in fact the opinion of certain figures such as Danny Hillis[17], who alerted the public about the dangers of an overload of the Internet [HIL 13]. Hillis asserts that massive use of the Internet could be harmful, since the network was not created for this kind of use. In the domain of connected health, a failure or a saturation of the network could have a serious impact. In addition, the Internet is an open network that lacks security; the many facts and reporting on backing as well as on the dark corners of the Internet (the dark web[18]) can only confirm this[19].

With the passage to "all-digital", consortiums and international organizations have been created to regulate the development of applications and technologies in various sectors. They set up inter-sectorial policies and attempt to regulate and align standards toward international development. Organizations for regulation and standardization (for example the European Telecommunications Standards Institute (ETSI), International Organization for Standardization (ISO), the Machine-to-Machine (oneM2M) group, Institute of Electrical and Electronics Engineers (IEEE, task force P2413) participate in the proliferation of applications and services by defining common rules so that all the actors are going in the same direction.

---

16 On February 3, 2011, the number of public IPv4 addresses was officially exhausted.

17 Danny Hillis is an American inventor, entrepreneur, author, engineer and mathematician. He is also the co-founder of the business Thinking Machines Corporation, known for having developed the Connection Machines supercomputer in 1983.

18 The Dark Web refers to web resources which are not accessible from a search engine (70– 75% of all web pages), but only via anonymization software such as Tor. In the absence of regulation, this part of the web is very free in the sense that you can find all kinds of content there, from the most illegal to the most traditional.

19 On March 17, 2016, the FBI made a public announcement warning owners and manufacturers of "connected" cars against the dangers of hacking with regards to this type of vehicle. The agency recommended vigilance against the vulnerability of vehicles to hacking and listed several "best practices": do not connect just anything to it, alert the manufacturer at the first sign of anomalies and carry out software updates.

The conception of devices and computer systems is shifting toward multifunction and concentration of services (for example telephony, television, web navigation). Since its debut in 1983[20], the mobile telephone has greatly evolved to integrate a large panel of functionalities ranging from the traditional functions of a telephone to Internet navigation to television. Users can read books, send electronic messages, listen to music, watch a movie and many other things as well. Devices became "all-purpose" tools. As an example, in *Everyware* (2007), Adam Greenfield highlights Japanese practices vis-à-vis the portable telephone where it has become an indispensable tool for most activities of daily life[21]. It is used for almost everything, from planning a meeting to searching for the nearest businesses, to the search for modes of transportation. The telephone as "universal remote" is being developed, as seen in Google's project entitled Physical Web[22].

2.2.2.5. Technology convergence

In addition to the digital convergence, which sees the fusion of support, transportation and content, we are also seeing a convergence of technologies, another factor involved in the expansion of the IoT. Going forward: "[...] telecommunications, information technology and media, sectors which originally worked largely independently from each other, are more and more integrated with each other. [...]" [PAP 07]. Appearing at the beginning of the 2000s, the convergence of technologies originally referred to the permeability between the borders of bioinformatics, information and communication technologies (ICT), cognitive sciences and microelectronics. The convergence of these new technologies is referred to under the acronym "NBIC": nanotechnologies, biotechnologies, computing (Big Data and the IoT) and cognitive sciences (artificial intelligence and robotics). At the heart of today's

---

20 The first commercial portable telephone commercial was launched on April 6, 1983, by Motorola in the United States. Dr. Martin Cooper, director of research and development at Motorola, was its inventor and as such he held a demonstration of the device on April 3, 1973.
21 Thesis 50, p. 117.
22 The Physical Web is a project launched by Google in 2014 with the goal of defining a universal standard to govern interactions between objects and smartphones. Google wants to set up a means by which the user no longer needs a multitude of applications dedicated to using objects. Thus, there is no longer a need for a specific application to look up bus schedules or pay a parking meter, since the protocol developed is universal and independent of any operating system, http://google.github.io/physical-web/cookbook/.

worldwide economy, NBIC industries are stimulating transhumanism[23] – a vast project intended to augment human performance on every front (physical, intellectual, moral and emotional). A first report developing this idea was edited in 2002, for the National Science Foundation [MIW 02], by Mihail C. Roco and William Sims Bainbridge, Converging Technologies for Improving Human Performance – Nanotechnology, Biotechnology, Information Technology and Cognitive Science.

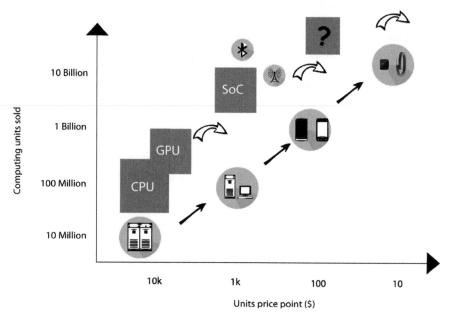

Figure 2.3. *Exponential evolution of components (from [RAN 14])*

The IoT was created from the simultaneous existence of technologies which make possible evolutions and innovations which have until now been theoretical or in the prototype stage. In an article published in the O'Reilly Radar on June 12, 2015, Susan Conant lists some of these factors [CON 15]:

---

23 In transhumanism, the improvement of the human species is confronted in one of two ways: a) in continuity, without renouncing his humanity; b) in a break with humanism, by technologically exceeding the limits of humanity (aging and death). With extropians and singularitarians, everything is possible. "If everything has become possible, is everything desirable?" [JOU 16].

– Moore's Law, formulated by Gordon Moore in 1965, stipulates that the complexity of entry-level microchips doubles every two years for an identical price[24]. Therefore, "speed", "power", "capacity", "clock rate" and many other properties of a computer system double every two years; over time, these systems become more and more performing for lower costs (see Figure 2.3);

– Metcalfe's Law, according to which the value of a network is proportional to the square number of its users. According to this law, the growing number of COs developed by businesses makes the size and popularity of the IoT grow;

– the multiplication of wireless communication technologies. Some examples include the popular Bluetooth[25], its variant Bluetooth LE (Low Energy)[26], its competitor ZigBee[27], WiFi soon omnipresent in all cities and towns[28], mobile telephony norms (for example GSM, EDGE, GPRS, LTE)[29] or NFC[30] and RFID[31]. A new form of wireless technologies called LPWAN (Low Power Wide Area Network) also emerged in this sector. This type of technology demonstrates low energy consumption for greater coverage, a model adapted to COs which must sometimes have an autonomy extending over several years and communicate over areas covering several kilometers;

– the progress of technological solutions related to batteries is opening new fields of application. For example, the autonomy of devices is increasing and the construction of electric cars is becoming relevant[32];

---

24 Moore's Law stimulates the production of bloatware and according to Wirth's law (1995): "programs slow down faster than the equipment speeds up."

25 Two-way communication standard with very short range.

26 A low-energy derivative of Bluetooth.

27 A low-energy communication protocol.

28 WiFi is in the process of becoming omnipresent in cities and towns. In 2015, the number of public access points was estimated at around 50 million (iPass – https://www.ipass.com/ press-releases/the-global-public-wi-fi-network-grows-to-50-million-worldwide-wi-fi-hotspots/).

29 Very-long-distance communications technology, infrastructures deployed by operators covering large wide geographic areas.

30 Near Field Communication, a means of communication with a very short range (a few centimeters).

31 Radio-frequency identification. Identifies objects using a tag that emits radio waves.

32 Marketed in the United States in 2012, the Tesla Model S electric car has an autonomy of 502 km, a record in this sector: http://www.usinenew.com/article/tesla-motors-comment-la-start-up-de-palo-alto-reinvente-l-automobile.N213178.

– miniaturization: the technologies used for the manufacture of sensors, actuators[33] and connected devices allowing their incorporation into smaller and smaller objects. Closely linked to Moore's Law, the process of miniaturization allows manufacturers to produce components at the micrometer or nanometer scale;

– Big Data[34] or massive data[35] represents the gigantic quantity of digital data generated by Internet users via their connected devices for personal or professional ends. Seen as an ensemble of tools and algorithms which make it possible to collect, store, process in real time, analyze and visualize very large quantities of data, Big Data is a global concept which relates to six variables (the 6V): volume (the quantity of the data generated[36]); variety (raw data, semi-structured or non-structured (for example: text, data from sensors, sound, video, browsing data, log files); complex data originating from very different sources such as social media, Machine to Machine interactions, mobile terminals, etc.); speed or velocity (the frequency with which the data is generated, captured and shared); veracity (the reliability and credibility of the data collected); value (the profit that can be earned from the use of Big Data); visualization (the possibility of rendering information comprehensible despite its volume, their variety and their constant evolution). Big data requires innovative technologies for the storing of non-structured data (for example Hadoop or NoSQL) and processing adapted in order to optimize time (for example MapReduce, Spark);

– Cloud Computing consists of platforms that will store, aggregate and analyze data from distant servers. If we follow the definition provided by the Office québécois de la langue française [OFF 15], Cloud Computing refers to a model in which computing resources are shared in the form of services

33 An actuator, unlike a sensor, refers to a device in an automated system that makes it possible to carry out concretely the action the computer commands. Thus, an acoustic speaker is an actuator because it emits a sound.

34 According to the archives of the ACM (Association for Computing Machinery) digital library, the expression "Big Data" seems to have appeared in October 1997, in scientific articles on the technological challenges faced in visualizing "large amounts of data."

35 As an official translation of the term "Big Data" the French Commission of Terminology and Neology recommends "mégadonnées": "structured or non-structured data whose very large volume requires adapted analytical tools." (*Le Journal Officiel* of August 22, 2014). We also find the expression "massive data".

36 By way of example, in 2011 the whole amount of data generated increased to 1.8 zettabytes, followed by 2.8 zettabytes in 2012 (CNRS, "The Big Data Revolution" *CNRS International Magazine*, January 2013).

and applications from servers connected to the Internet. The principal models of service offered with cloud computing are: Software as a Service (SaaS), Platform as a Service (PaaS), Infrastructure as a Service (IaaS). We also find Data as a Service (DaaS), Business Process as a Service (BPaaS), Network as a Service (NaaS), Desktop as a Service (DaaS), Storage as a Service (STaaS; for example, Dropbox, Google Drive, iCloud, Amazon Simple Storage Service, SkyDrive, Windows Live Mesh).

Cloud Computing applied to Big Data and the IoT allows the centralization of data and processing power. Cognitive analysis and machine learning techniques are some of the "tools" that allow the harnessing of these large volumes of data. The deepening of research into the cognitive domain and the improvement of learning techniques contribute to the elaboration of the IoT (see Figure 2.4).

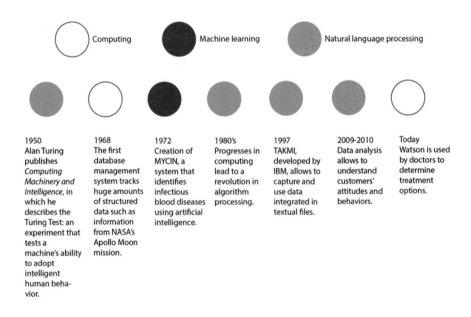

Computing   Machine learning   Natural language processing

1950	1968	1972	1980's	1997	2009-2010	Today
Alan Turing publishes *Computing Machinery and Intelligence*, in which he describes the Turing Test: an experiment that tests a machine's ability to adopt intelligent human behavior.	The first database management system tracks huge amounts of structured data such as information from NASA's Apollo Moon mission.	Creation of MYCIN, a system that identifies infectious blood diseases using artificial intelligence.	Progresses in computing lead to a revolution in algorithm processing.	TAKMI, developed by IBM, allows to capture and use data integrated in textual files.	Data analysis allows to understand customers' attitudes and behaviors.	Watson is used by doctors to determine treatment options.

Figure 2.4. *Advances in cognitive analysis (adapted from [BAT 14])*

There are obviously other factors ascribable to the development of COs and the IoT. Nevertheless, the elements mentioned above are among those having the most influence on the most recent innovations.

2.2.3. ... to the Internet of objects

The IoT sees the network and computing extended to every aspect of daily life. The Gartner Institute put the IoT at the top of the "Hype Cycle" curve [GAR 15] for the year 2015 meaning that this trend creates a large number of expectations and hopes (see Figure 2.5). Some people see it as the coming of a new era which would radically change our way of living and have an effect in all aspects of human life.

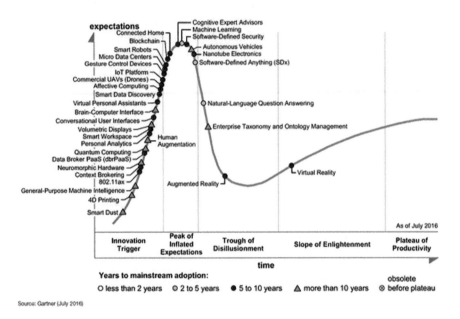

Figure 2.5. *The "Hype Cycle" of emerging technologies offered by Gartner in 2016*

Bit by bit, computing and networks are being incorporated in objects that surround us and more generally in our daily lives.

2.2.3.1. *Ubiquitous computing*

The term "ubiquitous computing" only appeared in 1999, but the idea had already been in progress for several years as various accomplishments such as the Internet Toaster clearly demonstrate. The idea of an interconnected world is not new and can be found under other designations such as "ubiquitous computing" "ambient technology" or "calm technology." In 1988, a scientist from Xerox PARC (Palo Alto Research Center) named

Mark Weiser, considered the father of this paradigm, theorized what he called ubiquitous computing. He explained that it involves the integration of computing tools into objects from daily life. In this way computing becomes omnipresent, systems and technologies will disappear, not physically but by being made invisible to humanity by merging with the environment and being incorporated into objects:

> "The most profound technologies are those that disappear. They weave themselves into the fabric of everyday life until they are indistinguishable from it […]." (Mark Weiser, *The Computer for the 21st Century*, p. 1, 1999)

In the same vein, in 1980, Ken Sakamura, Professor at the University of Tokyo, proposed the concept of an "ubiquitous network" [TRO 15]: computer network where all the objects that surround us integrate computing embedded with sensors and actuators connected to the same network allowing them to communicate between themselves. Dialogue and cooperation between devices would offer "smarter"[37] functionalities in the sense that one part of the operations carried out grows in autonomy.

These expressions denote an environment populated with COs and embedded computer systems – a phenomenon which is currently taking place since there are now more COs than people in the world (see Figure 2.6). A multitude of technologies and standards intended for identification appeared alongside COs and the services that ensue from them: sensors, connectivity, methods of communication or the network.

Like the Internet, as we know it today, the ecosystem of the IoT, if we can call it that, remains very heterogeneous. The term "ecosystem" is defined by the *Office québécois de la langue française* as follows:

> "A dynamic whole formed by living organisms and the non-living environment in which they evolve, their interaction making up the functional unity based on ecology." (Office québécois de la langue française, http://granddictionnaire.com, 2014)

---

37 Here, the term "smart" is preferred over "intelligent" since it would be clumsy and false to consider COs as "intelligent" in the same way as a person.

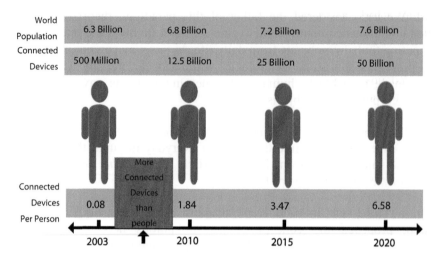

Figure 2.6. *Evolution of the number of connected devices in relation to the world's population (from [EVA 11])*

Transposed to the world of computing and the IoT, the ecosystem refers to all of the computer-based systems forming the environments in which they evolve and interact. Technologies and standards assure the interoperability and security of computing equipment and thus guarantee a certain stability and coherence. For example, the Internet is based on well-defined protocols such as TCP and IP and on international organizations such as the IETF[38] or the ISOC[39] which assure its governance. The same goes for the WWW which rests on the three fundamental standards of HTTP, URI web addresses and HTML. As for regulation, the W3C watches over the compatibility of technologies. The two form functional groups of, ecosystems. However, the IoT is in the development phase and it seems like that ecosystem remains to be built.

2.2.3.2. Connected objects

Before addressing the formal definitions of the IoT, we should define what we mean by "connected objects"[40]. Countless devices can become COs:

---

38 Internet Engineering Task Force, in charge of technical aspects and architecture.

39 Internet Society, responsible for the development of computer networks.

40 As Gilbert Simondon did regarding technological objects, the awareness of COs' mode of existence "must be carried out by philosophical thought" [SIM 12].

a lamp, a fork, a scale, a lock, a bed, an armchair or even a painting are objects that are potentially connectable.

We can consider a device a CO if its initial conception and purpose excluded any form of functionality calling on the concepts and notions related to the world of computing and the Internet network. An object such as a coffee machine or a lock was designed without the integration of computer systems or with a connection to the Internet. These objects are seeing their functions supplemented, just like John Romkey and Simon Hackett's toaster. A connection to the Internet was integrated into the simple toaster, allowing it to be controlled remotely. In contrast, according to this rule, the smartphone, the most connected of device would not be part of the family of COs.

Other features are then added to this. A CO is a device that interacts with the physical world without requiring human intervention. By their nature, COs have several constraints such as memory, bandwidth or energy consumption. Depending on the context in which it will be used, a device can be made to function for several years without discontinuity. Therefore, the energy consumption must be very low. This particularity has a notable impact on the speed and the frequency of the messages sent from the objects to service platforms.

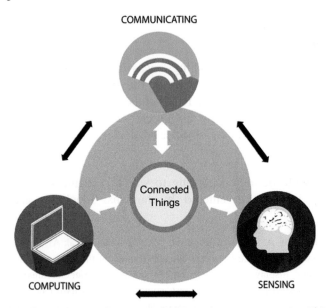

Figure 2.7. *Components of connected objects (source: www.smartthings.com)*

By their primal nature, these objects are physical, but also digital (see Figure 2.7). Here it is a physical object to which a certain form of artificial intelligence or AI (weak AI[41]) has been added so that the object can act in a "smarter" way [DVO 13], outside of logic that is irrelevant and not intuitive (for example a boiler that heats an empty building, a traffic light remaining "red" in the absence of any presence at opposite intersections, a microwave oven heating all dishes the same way, a bus stop from which it is impossible to follow the arrival of the bus, a shopping cart unable to count and add up the price of its content the work surface of fully-equipped kitchen unable to provide a catalogue of recipes, lights that don't automatically light up a dark room when a person enters it).

For example, today a red light only follows a programmer (or timer) that triggers a change after a precise amount of time has elapsed. The same stoplight integrating an AI would then be capable of determining on its own when the light should "turn green" or "turn red" depending on the number of cars and pedestrians present at an intersection. In addition, by adding a connection to the Internet and by generalizing the concept to all of the lights in a town, it would be possible to improve traffic in the town, by regulating it in a more relevant way: depending on current needs, the number of vehicles in the town, pollution levels or even schedules. An installation of this kind would contribute to making towns "cleaner" by reducing the pollution levels and would reduce the duration of motorists' trips.

2.2.3.3. *Definitions of the IoT*

The expression "Internet of Things" was invented by the Briton Kevin Ashton in 1999. He is the cofounder of the Auto-ID Center at MIT and participated in the creation of the RFID standard. The term was mentioned during a presentation for the company Procter & Gamble (P&G). Behind this expression there is a world of objects, of devices and of sensors that are interconnected [ASH 09].

He begins with the idea according to which computers and the Internet are based on information entered by humans. Machines are therefore

---

41 Two major classes of AI can be distinguished: strong AI and weak AI. Strong AI refers to a machine possessing all the characteristics and specificities of the human brain (for example, have self-awareness, display intelligent behavior and demonstrate feelings). This type of machine must be able to pass the Turing test, something which hasn't been produced yet. While a weak AI refers to a system with strong autonomy, for a very particular task. This autonomy gives the illusion of a certain form of intelligence, which is in fact a simulation.

dependent on our capacity to create data and transmit it, via the keyboard, by taking a photo, filming something or scanning barcodes. The problem, according to K. Ashton, is that Man is easily able to transmit ideas, but demonstrates difficulties apprehending environmental data on the (physical) things that surround him on his own. For this reason, computer systems have little data concerning the "things" of the physical world. Again, according to K. Ashton, the solution would be to allow machines to collect this data themselves. From this data and its analysis, it would be possible for example to better regulate energy consumption, reduce certain costs or even fight against waste more effectively.

Several versions of the IoT exist and so do many definitions emphasizing some points to the detriment of others. The concept is also often confused with machine to machine communication, the WoT or calm technology.

On May 27, 2015, the IEEE published a document whose goal is to obtain a definition of the IoT. The document explains that the IoT is present in diverse environments calling on different degrees of complexity according to the application scenarios set up. For this reason, the IEEE offers two definitions, the first defined environments and scenarios with a low level of complexity as follows:

> "An IoT [Internet of Things] is a network that connects uniquely identifiable 'Things' to the Internet. The 'Things' have sensing/actuation and potential programmability capabilities. Through the exploitation of unique identification and sensing, information about the 'Thing' can be collected and the state of the 'Thing' can be changed from anywhere, anytime, by anything." (IEEE, Towards a Definition of the Internet of things (IoT), p. 74, 2015)

The second definition, conversely, emphasizes more complex scenarios:

> "Internet of Things envisions a self-configuring, adaptive, complex network that interconnects 'Things' to the Internet through the use of standard communication protocols. The interconnected things have physical or virtual representation in the digital world, sensing/actuation capability, a programmability feature and are uniquely identifiable. The representation contains information including the things' identity, status, location or any other business, social, or privately relevant information. The things offer services, with

or without human intervention, through the exploitation of unique identification, data capture and communication, and actuation capability. The service is exploited through the use of intelligent interfaces and is made available anywhere, anytime, and for anything taking security into consideration." (ibid.)

In another study on the IoT, Pierre-Jean Benghozi, Sylvain Bureau and Françoise Massit-Folléa [BEN 08] offer a definition that crosses: "[…] purely technical approaches and techniques and approaches centered on usage […]" (p. 10). They define the IoT in the following way:

"[The IoT is] a network of networks which allows, via normalized and unified, electronic systems of identification and wireless mobile devices, to identify directly and without ambiguity digital entities and physical objects and in that way, be able to recover, store, transfer and process, without discontinuity between physical and virtual worlds, the data related to it." (p. 10)

In other words, the IoT is a network of networks because it is based on the Internet and it makes use of the concept of Internetting. This network is therefore made up of COs which are linked to it by means of wireless communication and which are uniquely identifiable. Connected devices produce data and transfer it on the network, cloud platforms[42] storing and then analyzing this data in order to provide a service or carry out an operation (see Figure 2.8).

The technical approach, as it is described in the study, suggests that the IoT is an: "[…] extension of the Internet naming system and indicates a convergence of digital identifiers […]" (ibid., p. 9). It involves extending the naming system specific to the Internet to COs, which allows internet users to navigate from site to site by using specific addresses for each resource. The identification methods used previously for the Internet network are applied to objects, computers, portable telephones or tablets. Thus, all terminals[43] are identifiable uniquely and automatically (without it being necessary to enter the identifier of the actor with which one would like to communicate).

---

42 Service platform hosted in the Cloud.

43 Terminal refers to one of the end points of a computer network. A work station connected to the Internet is a terminal, in the same way a lamp would be if it were also connected to a network.

Figure 2.8. *The steps of information (from [HOL 15])*

In addition to the technical approach, some definitions stress identity and usage. The omnipresence of computers, environments populated with COs and the services they provide, creates particular practices. The IoT would be a revolution connecting both people and objects, independent of time and location. Thus, this type of approach tends to personify objects by giving them a virtual identity and personality, especially because they are capable of communicating within the network. By being simultaneously real and virtual, objects become interfaces between these two worlds. This analysis is close to ubiquitous computing. By integrating into the environment, computer systems form a bridge between the physical world and the virtual world. A report created by several actors, including the European Commission and the EPSoSS (European Technology Platform on Smart Systems Integration), supports this approach. They place questions of functionalities and identities at the center of attention:

"Things having identities and virtual personalities operating in smart spaces using intelligent interfaces to connect and

communicate within social, environmental, and user contexts."
(Infso D.4 Networked Enterprise and RFID; Infso G.2 Micro
and Nanosystems and EPSoSS, Internet of things in 2020. A
roadmap for the future, p. 6, 2008)

Despite the plurality of definitions, we can identify several similarities.
Thus, the objects are linked to the Internet and have a presence on this
network. They are accessible and identifiable in the same way as a computer
or a smartphone. These devices act like gateways between the physical and
virtual worlds. Since the years 2010–2012 [MIC 15], a multitude of services
and products were introduced to transpose in real time the information
received from the environment in which they change into electronic data. It
is from this data, analyzed in Cloud platforms, that service providers develop
new functionalities (see Figure 2.9).

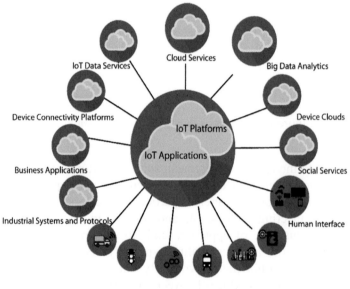

Figure 2.9. *Applications of the Internet of Things (adapted from [VER 15])*

Beginning with the idea that an infrastructure requires elements (a means of communication, a source of energy and a form of mobility), Jeremy Rifkin presents the IoT as the composition of an Internet of communications, an Internet of energy and an Internet of logistics [RIF 16]. The three components "[…] function together in a single system, by continually finding means of increasing efficiency and productivity to mobilize resources, produce and distribute goods and services and recycle waste. [...] Without communication, it is impossible to manage economic activity. Without energy, it is impossible to create information nor to fuel its transportation. Without logistics, it is impossible to move economic activity forward along the value chain" (p. 30). In addition to this composition, Jeremy Rifkin sees the IoT as a revolution making infrastructure intelligent and allowing a leap in productivity:

"[…] a revolution which will connect every machine, business, home and vehicle in an intelligent network made up of an Internet of communications, an Internet of energy and an Internet of logistics, all integrated into a single operating system." (ibid., p. 111)

2.3. Conclusion

From the creation of a World Wide Web connecting computers to its expansion into everyday objects, to the evolution of the web and the convergence of technologies, this chapter explains the historical and technological context which gave birth to the IoT and attempts to describe and define the concepts that revolve around this paradigm. The Internet connected billions of human beings, profoundly changing our usages and our ways of life, and the IoT is ready to do the same by connecting billions of objects. This accomplished, it has become an indispensable tool in our lives to share, create, modify and delete information. Normal objects become "connected" objects, an interface linking the physical world to the virtual world endowed with a certain form of intelligence to communicate with others, for example, thanks to sensors and embedded software. The CO can receive, contextualize, process and transmit data, by optimizing use and/or creating value. The following chapter describes the tools necessary for the conception and realization of the IoT.

2.4. Bibliography

[ASH 09] ASHTON K., "That 'Internet of things' Thing", *RFID Journal*, available at: http://www.rfidjournal.com/articles/view?4986, 2009.

[BAT 14] BATES C., "Cognitive Analytics", Deloitte, UK, available at: http://www2. deloitte.com/uk/en/pages/technology/articles/cognitive-analytics.html, 2014.

[BEN 08] BENGHOZI P.-J., BUREAU S., MASSIT-FOLLEA F., L'Internet des Objets. Quels enjeux pour les Européens?, Chaire Innovation & Régulation, Télécom ParisTech, 2008.

[CAR 98] THE CARNEGIE MELLON UNIVERSITY, "The 'Only' Coke Machine on the Internet", available at: https://www.cs.cmu.edu/~coke/ history_long.txt, 1998.

[CAR 15] CARTIER M., Le 21e siècle, Blog, 2015.

[CON 00] CONNOLLY D., "A Little History of the World Wide Web", available at: https://www.w3.org/History.html, 2000.

[CON 15] CONANT S., "The IoT will be as fundamental as the Internet itself – O'Reilly Radar", available at: http://radar.oreilly.com/2015/06/the-iot-will-be-as-fundamental-as-the-Internet-itself.html, 2015.

[DER 10] DE ROSNAY J., "Voyage vers le futur du web et la singularité", *TEDxParis 2010*, Paris, France, 2010.

[DVO 13] DVORSKY G., "How Much Longer Before Our First AI Catastrophe?", available at: http://io9.com/how-much-longer-before-our-first-ai-catastrophe-464043243, 2013.

[EVA 11] EVANS D., The Internet of things – How the next evolution of the Internet is changing everything, Livre Blanc, *CISCO*, 2011.

[GAR 15] GARTNER, "Gartner's 2015 Hype Cycle for Emerging Technologies Identifies the Computing Innovations That Organizations Should Monitor", available at: http://www.gartner.com/newsroom/id/3114217, 2015.

[GRE 07] GREENFIELD A., *Every[ware]*, FYP Editions, Limoges, 2007.

[HAU 03] HAUBEN R., "A Closer Look at The Controversy Over the Internet's Birthday! You Decide", available at: http://www.circleid.com/posts/a_closer_look_at_the_controversy_over_the_Internets_birthday_you_decide, 2003.

[HIL 13] HILLIS D., "The Internet Could Crash. We need a Plan B", *TED 2013*, Long Beach, United States, 2013.

[HOL 15] HOLDOWSKY J., MAHTO M., RAYNOR M.E. *et al.*, *Inside the Internet of things (IoT)*, Deloitte University Press, Westlake, 2015.

[JOU 16] JOUSSET-COUTURIER B., *Le transhumanisme*, Eyrolles, Paris, 2016.

[KUN 16] KUNKEL N., SOECHTIG S., MINIMAN J. *et al.*, *Augmented and Virtual Reality Go to Work: Seeing Business Through a Different Lens*, Deloitte University Press, available at: http://dupress.com/articles/augmented-and-virtual-reality/, 2016.

[KUR 07] KURZWEIL R., *Humanité 2.0: la bible du changement*, M21 Editions, Paris, 2007.

[MCC 01] MCCARTHY K., "World's first Webcam Coffee Pot to be Scrapped", available at: http://www.theregister.co.uk/2001/03/07/worlds_first_webcam_coffee_ pot/, 2001.

[MOR 14] MOROZOV E., *Pour tout résoudre, cliquez ici: l'aberration du solutionnisme technologique*, FYP, Limoges, 2014.

[OFF 15] OFFICE QUÉBÉCOIS DE LA LANGUE FRANÇAISE, Infonuagique – Le grand dictionnaire terminologique, available at: http://granddictionnaire.com/, 2015.

[PAP 07] PAPADAKIS S., "Technological Convergence: Opportunities and Challenges", *International Telecommunication Union*, available at: https://www. itu.int/osg/spu/youngminds/2007/essays/PapadakisSteliosYM2007.pdf, 2007.

[PAT 13] PATEL K., "Incremental Journey for World Wide Web: Introduced with Web 1.0 to recent Web 5.0 – a Survey Paper", *International Journal of Advanced Research in Computer Science and Software Engineering*, vol. 3, no. 10, 2013.

[RAN 14] RANADE P., THOMPSON S., SUVOLTA INC, "The $10 Price Point Will Drive the Next Wave of Computing", available at: http://electronicdesign.com/ embedded/10-price-point-will-drive-next-wave-computing, 2014.

[RIF 16] RIFKIN J., CHEMLA F., CHEMLA P., *La new nouvelle du coût marginal zéro: l'Internet des objects, l'émergence des communaux collaboratifs and l'éclipse du capitalisme*, Babel, Paris, 2016.

[ROC 02] ROCO M.C., BAINBRIDGE W.S., Converging Technologies for Improving Human Performance, Report, National Science Foundation, 2002.

[SIF 15] SIFAKIS J., "The Internet of things, une revolution à ne pas manquer. Prouver les programmes: pourquoi, quand, comment?", *Collège de France*, Paris, 2015.

[SIM 12] SIMONDON G., *Du mode d'existence des objects techniques*, Aubier, Paris, 2012.

[STE 15] STEWART W., "Internet Toaster, John Romkey, Simon Hackett", available at: http://www.livingInternet.com/i/ia_myths_toast.htm, 2015.

[TRI 11] TRIFA V.M., Building Blocks for a Participatory Web of Things, Doctoral thesis, ETH Zurich, 2011.

[TRO 15] TRON, Ken Sakamura has received International Telecommunication Union (ITU) 150 Award Blog, available at: http://www.tron.org/blog/2015/05/ topic0521/, 2015.

[VER 15] VERMESSAN O. *et al.*, "Internet of things beyond the Hype: Research, Innovation, Deployment", *Building the Hyperconnected Society: Internet of things Research and Innovation Value Chains, Ecosystems and Markets*, vol. 43, pp. 15–118, 2015.

[WCI 12] WCIT, "Convergence", *World Conference on International Telecommunications*, Dubai, December 3–14, 2012.

3

Introduction to the Technologies of the Ecosystem of the Internet of Things

By presenting the elements related to the context, the architecture and the protocols of the world of connected objects (CO), we will highlight the major scientific problems to be solved: precise identification of each object in a network, standardization and normalization of data transfer protocols, machine to machine (M2M) communication, encryption and security, legal status, and architectures of the Internet of Things (IoT).

Manufacturers build their products and services based on specific and sometimes proprietary architectures, thus rendering impossible the full comprehension of the product functioning. Interoperability between applications is limited by the will of companies to develop such proprietary models. As a result, Apple products are not compatible with Google products, Google products with Amazon products and so on.

The need for a common architecture is all the more important because it would make it possible to homogenize the conception of systems and favor compatibility and accessibility; it would also accelerate the development process and pave the way for new functionalities. Among the numerous architectures under development, models with common properties are beginning to emerge. We note the appearance of three-tier architecture and layered architecture which, in the end, are proven to be relatively close.

Chapter written by Ioan ROXIN and Aymeric BOUCHEREAU.

Finally, since 2015, a Request For Comments (RFC)[1] has been dedicated to interaction models in the IoT.

3.1. Architectures recommended by the Internet Architecture Board

In March 2015, the Internet Architecture Board (IAB) committe[2] edited the RFC 7452. It contains specifications dealing with the different conceivable architectures for the IoT. RFC 7452 presents four common interaction models between the actors of the IoT [TSC 15]:

– communication between objects;

– communication from objects to the Cloud;

– communication from objects to a gateway;

– from objects to back-end data sharing[3].

We shall briefly summarize these four models.

3.1.1. Communication between objects

The illustration below (see Figure 3.1) represents a wireless communication between two products from different manufacturers, a light bulb and a light switch. The transmission of information between the two devices is possible thanks to the integration of a wireless communication technology such as Bluetooth or ZigBee.

This type of model is very common for homes automation systems or for devices related to athletic activities (for example, step counters or heart rate monitors). The devices communicate through wireless channels by means of a network, most often based on IP (Internet Protocols) and on the Internet.

---

1 RFCs, literally "requests for comments" are a numbered series of official documents describing the technical aspects of the Internet or different computing equipment.

2 The IAB was entrusted by the Internet Society to oversee the development of the Internet. The organization is divided into task forces such as the IETF (Internet Engineering Task Force).

3 The term back-end refers to the non-visible part of a computer program. These are algorithms and other computer processing.

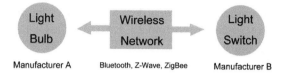

Figure 3.1. *Machine-to-machine communication (from The Internet of Things: an overview, Internet Society, 2015 [ROS 15])*

3.1.2. *Communication from objects to the Cloud*

In this type of communication, the data collected by sensors travels to service platforms via a network (most often the Internet).

Here (see Figure 3.2), the temperature and carbon monoxide sensors transmit the data they collect in real time to a specific platform located in the Cloud. Generally, the platform belongs to the manufacturer of the sensor and, because of that, these interactions only involve a single service provider. Consequently, there is no need to ensure interoperability with other manufacturers.

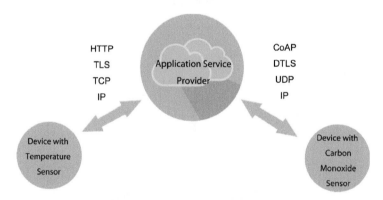

Figure 3.2. *Communication from devices to the Cloud (ibid.)*

Service providers are not immune to economic problems and so customers run the risk of seeing the products purchased become useless as a result of the bankruptcy of the manufacturer (and therefore the platforms developed around products). In response, transmissions are carried out on a network that supports IP to assure interoperability. This way, manufacturers authorize the development of third-party applications that can exploit their products. There are also various protocols and standards based on IP dedicated to communication between devices and service platforms such as CoAP (Constrained Application Protocol, RFC 7252); UDP (User Datagram Protocol); REST (Representational State Transfer) and HTTP (HyperText Transfer Protocol).

This type of architecture works thanks to the use of communication technologies such as Wi-Fi which is widely used and can cover several dozen meters. For example, this architecture model can be applied to a connected thermostat that will collect data then transmit it to an application in charge of storing and analyzing it. This processing allows the user to obtain details about their energy consumption.

3.1.3. *Communication from objects to a gateway*

Unlike the previous model, this one is more suitable for devices that cannot directly exploit the technologies defined by the IEEE standard[4] 802.11 (see Figure 3.3). Indeed, in some cases, an intermediary is necessary in order to make the connection between the sensors and the Cloud applications. Certain devices are sometimes incompatible with the Internet Protocol.

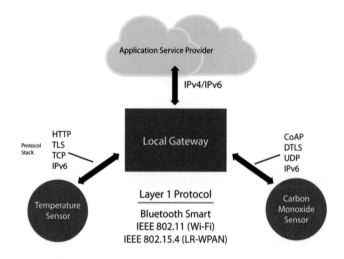

Figure 3.3. *Object-portal communication (ibid.)*

---

4 The Institute of Electrical and Electronics Engineers is a professional association responsible for the establishment of several standards in the area of electrical engineering.

Manufacturers use a "gateway" that retrieves information collected by the COs and then route it to the service platform. For example, the smartphone is a gateway between the smart wristband and the online application.

The advantage of this type of architecture is that it allows the addition of devices not compatible with protocols favoring interoperability such as IP, in a system adapted to this type of technology. However, this approach is costly, since it requires the development of additional applications for the gateway.

3.1.4. *From objects to back-end data sharing*

The model presented here (see Figure 3.4) address the problem of data sharing between service providers.

Indeed, most of the time the data generated by COs are sent to one single platform, thus preventing the exploitation of data by third-party providers and applications.

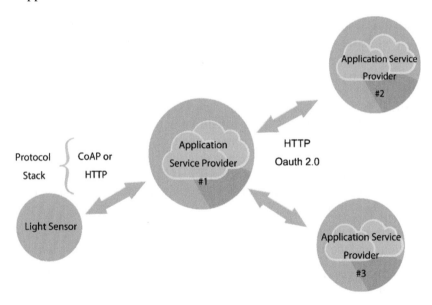

Figure 3.4. *Architecture with back-end data sharing (ibid.)*

Nevertheless, some manufacturers are developing an API (Application Programming Interface), paving the way for the exploitation of the data aggregated by external manufacturers. This is the "programmable web" concept. The platforms set up APIs, most often a REST web service, which allows access to all, or a part, of the information collected by the service.

By way of example, the data collected by an activity tracker could be used by a service specializing in behavioral analysis. Using the data created by movements, third party algorithms can for example dispense advice on physical activity.

3.2. Three-tier architecture

A large number of groups have embarked on the development of a standard architecture for the IoT. Nevertheless, these efforts are often concentrated on specific applications [WG4 16]. For example, Air Force Enterprise Architecture Framework is specifically interested in Air Force IT systems. Similarly, the French project AGATE (*Atelier de gestion de l'architecture des systems d'information and de communication*) is only intended for the French Armory.

The standardization organization IEEE Standards Association (IEEE-SA) has also embarked on this path, but with the view of developing a model adaptable to all application contexts. That is how the IEEE P2413 [IEE 16a] task force was born.

IEEE P2413 provides a model and specifications making possible to go beyond the existing "barriers" between different domains. The fact is that the needs for setting up the IoT vary from one sector to another. Thus, more than any other, attention and rigor are required in the medical field when it comes to the security and reliability aspects [IEE 16b]. This is less the case for leisure dedicated COs. Faced with this problem, the task force has settled a the following objectives:

– establish a standard model to define the relationships, the types of interactions and the architectural elements common to diverse domains;

– develop a standard architecture that is compatible, whatever the field of application.

According to the IEEE-SA the work carried out should be finished in the course of 2016. For now, the architecture considered by the IEEE P2413 task force is the following (see Figure 3.5):

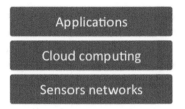

Figure 3.5. *Three-tier architecture*

The architecture is made up of three levels. The lowest level corresponds to sensors and the communication between them (Machine-to-Machine). It is the network of sensors and actuators. Located at the base of the model, it generates the data upon which the offer of services is built.

As for the second level, it is occupied by Cloud Computing, the service platforms to which the data are intended. This level makes the link between the sensors networks, and the network of platforms and data-processing programs. This is where the data is carried for the purposes of storage, aggregation and analysis.

The third and final level concerns applications and services offered to customers. Thanks to data coming from the networks of sensors and its analysis by specific computer programs located in the Cloud, businesses can expand their services offer: environmental surveillance, medical surveillance, optimization of energy consumption, etc.

This model has the particularity of giving Cloud computing an important role. The model is called "Cloud-centric"[5] because it rests, in a large part, on the Cloud. From the IEEE, this vision is not surprising since in its IoT definition the organization had already established Cloud Computing as a critical element.

3.2.1. *Layered architecture*

In addition to the three-tier architecture developed by the IEEE P2413 task force, there is another one, structured in "layers." This architecture is often used to describe the structure and existing relationships between the different IoT actors.

---

5 "Cloud-centric" refers to a concept centered on or based on cloud computing.

This architecture takes the form of superimposed layers, following the model thought up by the IEEE-SA. No official nor universal version exists, but there is an average of six to eight different layers. Each one of these layers represents an activity vital to the functioning of the IoT, with one or several actors.

Through these different levels, we can also distinguish the path taken by the data resulting from the monitoring of the physical world (the environment).

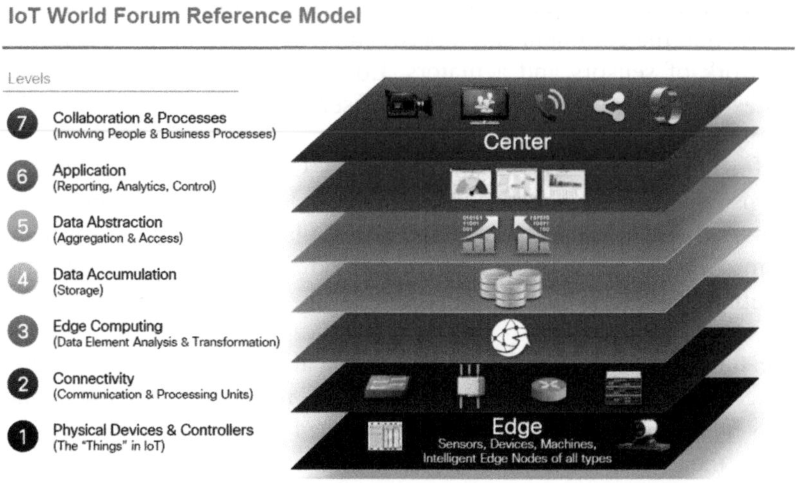

IoT World Forum Reference Model

Figure 3.6. *The different layers of the IoT (from Cisco [GRE 14])*

For example, the American company Cisco, specialized in networks, takes an interest in the IoT, especially in its connectivity aspect. In the presentation "Building the Internet of Things" dating from October 2013, Jim Green, the Chief Technology officer, presented the model his business intends for the IoT. Here we can find the seven layered architecture illustrated above (see Figure 3.6).

The details of the seven layers defined by Cisco are the following:

– the physical layer is the first of the seven. This layer is made up of devices and connected objects such as: temperature sensors, connected watches, physical activity trackers, connected lights, smart coffee machines, smart glasses, etc. The data originates here thanks to the "recordings" made by the CO integrated sensors;

– the next layer deals with the connectivity and communication processes deployed between the actors of the physical layer (the lowest layer) and those of the higher layers. Related to the previous model, this part is the equivalent of the "Networks of sensors" level described by the IEEE P2413;

– "Edge Computing" is a technique used by certain manufacturers to, among other things, reduce communication distances, improve transmission bit rates and reduce costs. It reduces distances between products, customers and service platforms by bringing together a part of the infrastructure and processing power. This technique resembles the "gateway" concept mentioned previously in the section, "Communication from devices to a gateway." Typically, the use of a mobile phone as an intermediary interface between the Cloud service and the CO can be considered Edge Computing. In relation to the model presented by the IEEE P2413, this layer is located at the border between the Networks of sensors and Cloud Computing levels;

– data storage. After being created in the physical layer and routed to the Cloud, the data is then stored in the data centers of service providers;

– after storage, comes the aggregation of all of the data collected. This step consists of grouping data according to specific classifications, in order to create consistent groups of data;

– the next layer concerns applications and algorithms (deep learning) which will analyze the data. This layer, with the two preceding ones (storage and aggregation), corresponds to the Cloud Computing level described by the IEEE task force;

– the last layer is that of the services and applications proposed to users. Algorithms analyze the data to create value, for example statistics provided by the applications accompanying athletic devices or advice for energy optimization proposed to smart thermostats' users. This step corresponds to the Applications level in the model conceived by the IEEE.

3.3. Steps and technologies in the ecosystem of the IoT

The "everything connected" tendency is permeating every aspect of our everyday lives: forks, shoes, cameras, tee-shirts, cars, books, coffee machines, bracelets, watches, tables, pens, speakers, televisions, notebooks or even glasses. All domains are involved, whether it is health, the home, modes of transportation, infrastructure and even entire towns. The problem that arises is the following: how is this going to work?

Numerous studies attempt to imagine the computing and Internet landscape of 2020. The COs number being one of the recurring themes, this indicator gives an idea of the possible expansion and growth of the IoT. Estimations are numerous with as many different numbers as studies made. For example, Intel forecasts 200 billion objects in 2020 [INT 16] while Cisco estimates the number of COs at 50 billion [CIS 16]. Nevertheless, all of these studies agree on the fact that billions and billions of objects will invade our environment by 2020.

Figure 3.7. *Setting up the IoT*

Admittedly, COs are at the heart of the IoT, but still it is necessary to connect all these devices, enable them to exchange information and interact, within the same network. In addition to the multiplication of COs, the setting up of the IoT goes through several other steps, illustrated in Figure 3.7:

1) Identification. Knowing how to precisely determine which object is connected to what, in what way and in which location. All of this is done remotely, in such a way that in a 50-storey building, the electrician in charge of repairing the electrical installations must be able to identify a defective light bulb with precision (for example floor, room, position, number), from his workstation. This requires a complete naming system capable of supporting the future growth of the number of terminals;

2) Capture. In order for COs to fulfill their role of bridge between the physical and virtual worlds, sensors are indispensable. Their multiplication, miniaturization and integration into the environment has proven to be necessary. They represent the "sensory organs" of objects. Located at the bottom of the IoT's chain of actors, sensors are the data source that feeds applications and services;

3) Connection. Linking objects with each other so that they can exchange data in a more autonomous manner. For example, in a house we would like objects to make decisions and set up operations that go in the same direction and that they act in concert. The lamps and the shutters need to communicate with each other to coordinate their actions. In this way, when the shutters sense that night is coming, they close automatically, but also sent a signal to the lamps that will then light up;

4) Integration. Connecting COs to the virtual world with the help of a wireless communication method. They are identifiable, they capture data and communicate with each other, but it is also necessary for them to be able to share their data with service platforms. To do so, each device integrates a communication technology (for example Bluetooth, ZigBee, NFC, Wi-Fi) which will allow the information transfer;

5) Networking. Users want to be able to interact remotely with their objects while the providers want to collect the data generated, that is often the service basis. Consequently, the objects are connected to a single network linking providers and their Cloud platforms and thus making them capable of being piloted remotely. It goes without saying that the Internet is the most appropriate for this task.

The setting up of the IoT therefore goes through the following steps (see Table 3.1): identification, sensors setup, object interconnection, integration into the virtual world, and network connection.

Identify	Capture	Connect	Integrate	Network
Making possible the identification of each connected element.	Implementation of devices bringing the real and virtual worlds closer. The objects basic functions (the temperature sensor for the thermometer for example).	Establish a connection between the objects so they can communicate and exchange data.	Use a communication means connecting objects to the virtual world.	Linking objects and their data to the computing world via a network (the Internet, for example).
IPv4, IPv6, 6LoWPAN	MEMS, RF MEMS, NEMS	SigFox, LoRa	RFID, NFC, Bluetooth, Bluetooth LE, ZigBee, Wi-Fi, cellular networks	CoAP, MQTT, AllJoyn, REST HTTP

Table 3.1. *Steps and technologies to set up the IoT*

The following parts will describe conceivable technical solutions for each step.

3.3.1. *Identifying*

We have billions of objects, and the first step to be taken, in order to reach the IoT as it has been defined above, is their identification. Indeed, all of the objects must be individually identifiable. The smart coffee machine in your office must have an unique id, so that it is identifiable through a vast network of terminals.

We must be able to identify each object in a network. In other words, at the scale of a smart house, already having a large number of objects (for example the coffee machine, refrigerator, oven, bed, garage, shutters, boiler, locks) all of them have to be linked via a single network. It must be possible to precisely identify the coffee machine or one of the kitchen lights, in order to replace it, for example.

This identification stage has already been thought of and solved by the Internet protocol where it is necessary to identify each computer. IP addresses are used to do this. They function similarly to addresses used by postal services to deliver the mail. If you would like to send a message to another computer connected to the Internet, it must have an address (an IP). Using the address, the services redirect the message to the recipient computer.

3.3.1.1. *From IPv4 to IPv6*

Since the 1980s (and still today), the large majority of computer systems use the IPv4 protocol. Although it was completely satisfactory at the beginning of the Internet, when the number of Internet-connected machines was low, it has turned out to be very limited, and incompatible with the IoT vision. Indeed, defined by the RFC 791 in 1981 [POS 81], the IPv4 is coded on 32 bits, which means that we can allocate only 2^{32} unique addresses (equal to 4,294,967,296). Yet, the number of terminals connected to a network has not stopped growing, all of the addresses are as of now allocated[6]. Therefore, it is not imaginable to apply this protocol to the IoT. To solve this problem, a new protocol called IPv6 was established.

IPv6 is coded on 128 bits which allows to create 2^{128} unique addresses (equal to 3.4×10^{38}). Thus, the protocol offers sufficient room for growth. The IoT involves a profusion of connected devices through a single network.

---

6 On February 3, 2011, the number of public IPv4 addresses officially reached the saturation point.

So, because of the large number of addresses that it offers, IPv6 seems to be the ideal candidate, allowing a unique identification for each object. With IPv6, it is estimated that several trillion addresses will correspond to each human being on Earth [IEE 07]. Despite the billions and billions of COs which are expected to appear by 2020, the IPv6 will be far from saturated.

3.3.1.2. *IPv6 Low power Wireless Personal Area Network (6LoWPAN)*

6LoWPAN is the equivalent of the IPv6 protocol adapted to low-consumption devices. IPv6 solves IPv4's problem by allowing the attribution of addresses to billions of COs. Nevertheless, some difficulties remain for this protocol to be integrated into sensors. The header size of the emitted packet is too large for this type of device. The calculations that must be made are complex and require high energy consumption to process it [MUL 07].

In this context, in 2005, the IETF[7] created the 6LoWPAN task force. Its goal is to solve the problems related to the implementation of IP in sensors. After a few trials and the publication of an RFC, in September 2007, the task force published RFC 4944 which finally allowed devices using 6LoWPAN technology to connect to the Internet. The technology compresses the headers of IPv6 packets so that it is feasible to implement them on different devices.

6LoWPAN is based on the communication protocol defined by the IEEE: 802.15.4. The latter is dedicated to wireless network LR WPAN (Low Rate Wireless Personal Area Network). In other words, it deals with devices combining both a short range, low consumption and low bit rate. Its purpose is to interconnect systems with few resources, such as sensors.

The development of a communication protocol via IP offers numerous advantages. The sensors are interoperable with diverse communication technologies such as Wi-Fi, Ethernet or cellphone networks. In addition the devices benefit from the network's security tools functioning with IP, naming systems and more generally from all of the means implemented in the past several years to ensure the IP sustainability.

---

7 The Internet Engineering Task Force is an organization open to everyone who participates in the conception of Internet standards.

3.3.2. *Capturing*

As we have just seen, it is necessary to precisely identify each COs in a single network in order to access them, control them, run programs, carry out maintenance or modify behavior. All of this is done remotely, independently of the user's geographical location. After the identification step comes the data capture, namely the setting up of sensors to convert analogic information to digital.

In order to function, COs must have sensors and/or actuators. Without that, they are no longer able to carry out the smallest tasks and become blind in a manner of speaking. The objects must have sensors to record events relating to the environment and the world that surrounds them. As for the actuators, they allow COs to act on the physical work. They are the link between the electronic command and the physical action.

Hence your coffee machine must, for example, have sensors allowing it to follow the level of the coffee supply. Or the coffee machine can be endowed with temperature sensors and actuators in order to heat up the cup.

If we would like COs to make our everyday lives easier and free us from unrewarding and unpleasant tasks, it is necessary for them, just like us, to have "senses": sight; touch; hearing; smell and taste. Sensors are therefore the COs "senses" the sensory organs. Many types exist: acoustic, pressure, movement, acceleration, light or temperature sensors.

3.3.2.1. *Micro-electromechanical systems (MEMS)*

The conception of sensors and actuators use MEMS technology (Microelectromechanical systems) [CIV 12]. It is a small system, one at the micrometer scale, made up of mechanical elements using electricity as energy. Developed at the beginning of the 1970s and marketed several years later, this microelectronic system is present in a quantity of everyday objects. It makes it possible to transform physical phenomena into electrical signals, thanks to the combination of computing, electronic, chemical, mechanical and optical technologies. These microsystems serve as an interface between physical phenomena, the environment that surrounds us and the electronic world (signals).

MEMS are mostly made up of transducers that "capture" the world that surrounds them, in order to subsequently transform it into electrical signals. Without exception, a transducer focuses only on a single physical phenomenon. Thanks to its microsystems, a connected thermometer can capture the ambient temperature, lamps vary in intensity automatically depending on ambient brightness, security devices can detect the presence of individuals or the plumbing system can detect possible water leaks.

There is also a MEMS derivative called RF MEMS, standing for "radio frequency." They resemble each other but the latter is dedicated to devices integrating communication with radio frequencies. Developed at the beginning of the 1980s and put aside for a long time, it is now used in antennas.

3.3.2.2. Miniaturization

In addition to MEMS technology, miniaturization has played a dominant role in the sensors proliferation. Indeed, sensors as well as other devices such as batteries or components used for wireless communication, are becoming minuscule. This is a trend that pushes businesses to create electronic products at smaller and smaller scales. As a result, they sometimes end up on the scale of a micrometer or a nanometer. Today only used in laboratories, NEMS technology is a miniaturization, at the scale of a nanometer, of its big brother MEMS. NEMS stands for: Nanoelectromechanical systems. Related to Moore's law, the computer systems' production tends towards miniaturization while gaining in performance, at a low cost. These are, among other things, the principal factors responsible for the development of embedded computing in any imaginable or possible object.

3.3.3. Connecting

From this point forward, our COs are identifiable in a unique way via IPv6 protocol and are equipped with sensors and actuators functioning with MEMS technology, allowing them to transpose the environment in electrical signals. COs produce data and now need to communicate with each other and exchange information.

The IoT means the birth of a world where computing is omnipresent and pervasive due to a multitude of COs and computer systems. This is also the case with networks: today there is connectivity and networks everywhere around us. It is through networks that objects can communicate and carry out actions cooperatively for a better user experience. In order for there to be a certain coherence in the behavior of all of these COs, it is essential to interconnect them (see Figure 3.8).

Acting autonomously, without any other actors' help (object or human), these COs provide only minimal improvements. Hence the interest in having "connected" objects that can communicate with each other. Thus, it would be nice if your smart home recognized you when you come up to your front door and opened that door for you. In doing so, it would send a message indicating your arrival to the heater as well as the lights. The message would launch specific programs for light and heating. Another possible chain reaction would be an alarm clock that detects when a person wakes up and transmits the information to the coffee machine which would turn on immediately. Therefore, there is little interest in objects acting only on their own. A coffee machine that cannot communicate with the alarm clock or the front door will be quickly limited in terms of functionality.

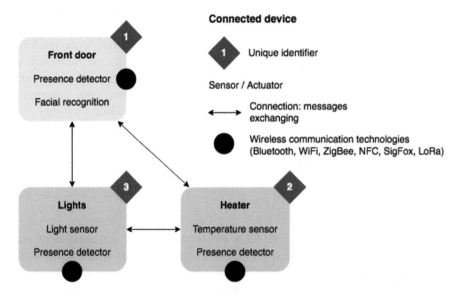

Figure 3.8. *Communication between COs*

The interconnected objects are being provided with new functions that will make their use much more relevant. Right now, objects developed by major companies such as GAFA (Google, Apple, Facebook and Amazon) do not communicate with each other, or on a minimal and negligible scale. It is a fragmented IoT where each company has created its own ecosystem, namely a hermetic ecosystem, closed to external actors, just like the web 2.0 where the service platforms have created closed spaces[8].

Because of the multitude of ecosystems there is no true ecosystem: a space in which objects speak the same language, communicate, and where a form of cohesion and cohabitation emerges.

3.3.3.1. *Machine-to-machine communication*

This operation mode, from object to object, is called M2M. It is the implementation of a means for establishing communication between devices with the same application, in the same context, via wired or wireless communication technologies. M2M allows costs reduction, increased productivity and both safer and more secure processes. Machine-to-Machine communication is an integral part of the IoT. It defines interaction modes between objects. In this paradigm, interaction is no longer just done within a Man-Machine framework, but also at the Machine-Machine level. To concretely define it, a M2M system is made up of devices able to capture data coming from external events, a network through which these devices communicate as well as a Cloud platform to collect, store and analyze thee data gathered for the purpose of carrying out operations [HOL 14].

Machine-to-machine communication is mostly carried out on LPWAN network infrastructure (Low Power Wide Area Network). This type of network is particularly adapted to low-consumption devices requiring coverage over large distances. LPWAN favors small volume communications at low bit rates and over relatively long time periods. The sensors and actuators used by our applications need to function for several years, sometimes as much as a dozen. In addition, the distance separating the application of the sensors can be counted in kilometers. Nevertheless, these devices are intended to emit and receive data for years. For this reason, technologies capable of ensuring low-consumption communication and wide area coverage were developed.

3.3.3.2. *SigFox*

SigFox is a French company created in 2009 which has developed a M2M LPWAN-type network. Its network is energy-efficient and has a low

---

8 For example, Facebook offers several services that are only accessible to those who have a user account. All its services and content is are protected behind authentication mechanisms and inaccessible to other actors on the web. The data generated is not open.

bit rate [WAT 14]. The network runs on UNB (Ultra Narrow Band) technology, which offers transmissions of several dozen hertz. In comparison, the signals emitted over the GSM network reach hundreds of kHz, and in some cases MHz.

SigFox uses ISM frequency bands (available globally and without a license, therefore free) for communications. The transmissions are limited (by SigFox) to 140 messages per day and are bi-directional, that is to say that a terminal can receive and emit information. The size of the messages sent is variable but cannot exceed 12 bytes.

Finally, SigFox is a proprietary network, which means that businesses must pay a subscription to benefit from the service. In return, the antenna installation, infrastructure and the network management are carried out by the company SigFox itself.

3.3.3.3. *Long Range Wide Area Network (LoRaWAN)*

The LoRaWAN protocol is a SigFox competitor. Just like SigFox, it works by radio and allows communications at a low bit rate over large distances.

This technology was developed in 2012 by the company Semtech under the name LoRa Alliance. Unlike SigFox, LoRaWAN is an open technology, which means that any business can develop its own LoRa network. In return, businesses must establish the necessary infrastructure themselves. In other words, COs that are part of the network must integrate a LoRa chip and the antenna installation (relays) linked to the Internet is in charge of the business developing a M2M LoRa network.

Just like SigFox, LoRa also uses ISM frequency bands and allows bidirectional communication. The transmissions can go from 15 to 20 km and the devices have an autonomy of a dozen years. LoRa allows a transfer bit rate between 0.3 kbps and 5 kbps [POO 15].

3.3.4. *Integrating*

At this point, we now have uniquely identified COs thanks to IPv6. MEMS technology allows a multitude of sensors and actuators serving as sensory organs so they can capture data and act on the physical world. The CO is capable of collecting data and sharing it with its counterparts, which is known as Machine-to-Machine dialogue. It is now important that devices have access to a communication means to interact with service platforms and users.

Each CO must integrate a communication technology allowing it to externalize its data and transpose them first from the physical world into electrical signals, with the help of transducers, then to convert the signals into computer data. There are numerous solutions capable of making the connection between objects and the virtual world, each one attempting to solve the problems raised by the diverse situations with which the devices will be confronted (see Figure 3.9).

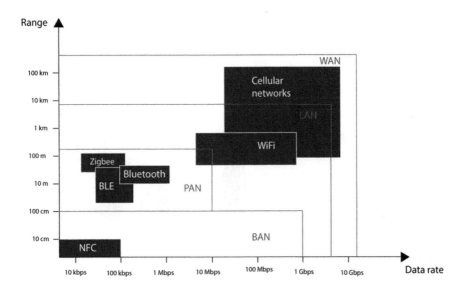

Figure 3.9. *Wireless communication technologies landscape*

The characteristics required for the CO connectivity depend on the application context. In other words, the future use of the product influences the choice of communication means. A smart wristband can easily settle with a range of a few centimeters while, on the other hand, a traffic light requires a range of up to several kilometers. The same goes for energy consumption, because COs are not directly linked to an energy source which requires them to function autonomously and, in some cases, they have to remain lit for a dozen years. A smart lock must remain on 24 hours a day, seven days a week and cannot be connected to a power source nor recharged every day. A power failure would have serious consequences (burglary, for example). The following factors are among the most affected by the future use of COs:

– energy consumption;

– communication speed;

– transmission quality;

– cost;

– security;

– range.

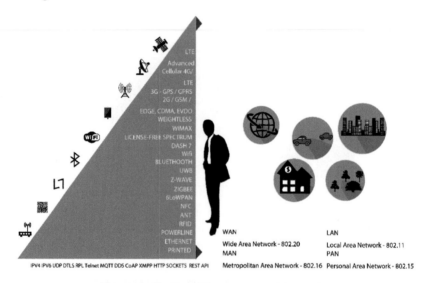

Figure 3.10. *The different types of networks
(adapted from Postscapes.com, 2016 [POS 16])*

It is possible to categorize the technologies, directly according to signal range and indirectly according to their future use (see Figure 3.10). We can distinguish wireless technologies BAN (Body Area Network), PAN (Personal Area Network), LAN (Local Area Network) and WAN (Wide Area Network)[9].

9 There is also the MAN (Metropolitan Area Network) category. Nevertheless, it includes very few technologies (for example WiMAX).

3.3.4.1. *Body Area Network (BAN)*

BAN is a category made up of wireless network technologies designed to interconnect devices, on, around and in the human body. These miniature systems are equipped with sensors and actuators that are able to measure specific human traits (for example the number of steps, heart rate, blood pressure) and act on them. They communicate via the same wireless network.

IEEE standard 802.15.6 was developed, as its name indicates, by the Institute of Electrical and Electronics Engineers. It is an extension of another standard known as 802.15[10] on wireless technologies intended for the higher PAN category networks. IEEE 802.15.6 provides BAN networks specifications. It defines a standard model for this type of wireless network implementation [KWA 10]. A BAN network is characterized by the following points [ASA 14]:

– low energy consumption (autonomy of several days or months);

– a high bit rate of data (higher than 1 Gbps);

– quality of transmission (little data loss);

– short range (up to 10 m).

The main application areas are health and sports. In the medical field, this type of network is particularly useful to follow in real time changes in the patient's vital signs. Sensors are becoming more and more efficient, and it is possible to detect vital signs such as heart rate, electrocardiogram or blood pressure with precision. Such devices allow hospitals to detect imminent heart problems of monitored patients.

For the same reasons, this type of network has proven to be very useful for athletes who can follow and analyze their performance. Athletes and trainers can see their speed, heart rate, the number of kilometers traveled, the number of calories burned or even blood pressure.

In the landscape of technologies that are able to integrate a BAN network, we can include RFID and NFC.

3.3.4.1.1. Radio Frequency Identification (RFID)

RFID, or radio-identification, is an automatic identification technology that uses radio frequency to identify objects. Equipped with a chip or another

---

10 At the higher level, the IEEE 802 oversees the standardization of wireless technologies.

similar device, they can transmit information via a radio antenna to a reader designed for this purpose [OFF 08].

In 1983, the first patent using the abbreviation "RFID" was filed by the inventor Charles Walton, considered the father of this technology. In 1999, MIT founded the Auto-ID Center where the inventor of the term "Internet of Things" Kevin Ashton, worked. This research center specialized in automatic identification. In 2004, the center was transformed into EPCGlobal. An organization responsible for the promotion of the EPC standard which is an extension of traditional bar codes [CNR 16].

Devices subject to radio-identification integrate a "radio identification tag." An adhesive component which includes the radio antenna as well as the memory chip. The chip contains an identification number that cannot be modified once it has been written[11]. Certain chips have a storage space for complementary information. The radio identification tag is called "passive" which means that, except the energy provided by the reading device, it doesn't use any other energy source. The reading can be done at a distance of up to 200 meters from case to case.

The applications are diverse, RFID can serve as a badge for accessing a building, for a bus ticket or to ensure the traceability of a business's products.

3.3.4.1.2. Near Field Communication (NFC)

NFC or near field communication is a wireless communication means with a very short range, in the order of a few centimeters (around 10 cm). Thanks to NFC, two devices can exchange information at bit rates of 106, 212 or 424 kbps [CUR 12]. Three modes of communication are possible:

– card emulation. The device use is the equivalent of using a card or a badge. It is passive, waiting to be read;

– reader. In contrast to the previous setting, the device is now active. It can read information originated from electronic tags;

– peer-to-peer. The two devices exchange data.

---

11 This is the principle of WORM (Write Once Read Multiple), a storage technique that authorizes a single writing, but several readings.

The first use of NFC goes back to the year 1997. The toy company Hasbro marketed the Star Wars CommTech Reader accompanied by several figurines representing the main characters. It used an older version of the NFC called CommTech. When one of the figurines was placed (more precisely the bases of the figurines, which contained an electronic chip) on the CommTech Reader, an audio message was emitted. The electronic chip in the figurine transmitted a message containing the soundtrack to the reader which then emitted sounds [GIL 12]. It was only in 2003 that the technology was formalized as a standard by the ISO/IEC (the International Organization for Standardization and the International Electrotechnical Commission) and the ECMA (European Computer Manufacturers Association).

This solution is currently used and implemented in applications such as contactless payment, vehicle startup access, profiles exchange between two users on the same social network, access to building automated functions or the reading of information about a product.

3.3.4.2. *Personal Area Network (PAN)*

Just like BAN, PAN is a short-range wireless network (a dozen meters). It is based on the IEEE standard 802.15 and is adapted to communications between peripheral devices and computers on short distance transmissions. It ensures the connection between mouse, keyboard, printer, speakers, tablet, smartphone and computer. More generally, PAN replaces all the wired connections that would otherwise exist around the computer and the mobile phone.

IEEE 802.15 defines a certain number of standards for PANs, including Bluetooth, Bluetooth LE and ZigBee technologies.

3.3.4.2.1. Bluetooth

As a popular short-range communication standard, Bluetooth is being implemented in a large number of products: smartphones, keyboards, headsets, smart wristbands, smart watches and wireless mouses.

Bluetooth was originally conceived to let telephones communicate without a wired connection. In the 1990s, several major companies (for example Intel, Nokia, Ericsson) tried to develop a technology of this kind. In 1996, these businesses decided to adopt a common technology and a common name to avoid fragmentation. Thus, in 1998 the technology was launched under the name "Bluetooth"[12] [KAR 16].

Bluetooth allows two-way communication by using radio waves. The transmissions have a very short range and there are three classes of Bluetooth, each one with a different range, up to several meters (see Table 3.2).

Class	Range
1	100 meters
2	10 to 20 meters
3	Several meters

Table 3.2. *Range of each Bluetooth class*

In addition to the relatively short distance, the communications have the particularity of consuming little energy and being inexpensive. These are the features that have made Bluetooth popular and have favored an almost systematic implementation in smartphones. As for the bit rate, it can reach up to 720 kbps [BLU 16].

3.3.4.2.2. Bluetooth Low Energy

Bluetooth LE (Low Energy) or Bluetooth Smart is a variant of the Bluetooth technology presented in the previous section. Created by Nokia, this version has the characteristic of consuming much less energy. Since energy consumption is a strong constraint in the IoT, Bluetooth LE could be the ideal solution to ensure the CO connectivity within a PAN network.

Nokia developed this technology in 2006 under the name Wibree [GRA 06], with the purpose of creating a Bluetooth equivalent, but with noticeably lower energy consumption and cost. In 2007, the members of

---

12 The name "Bluetooth" was proposed by Jim Kardach (an Intel engineer). The idea was given to him indirectly by Sven Mattisson (an engineer at Ericsson). The latter told him about the book *Longships* (by Frans G. Bengtsson) which was about, among others, the Danish king Harald Bluetooth (the English version of the name). The king was known for having unified and Christianized Denmark. In the same way that in his time King Harold unified Denmark and Norway, Bluetooth would unify two devices.

Bluetooth SIG have agreed to integrate Wibree into the specifications of the new Bluetooth version 4.0; thus becoming Ultra low power Bluetooth technology. It was only in 2010 that Wibree was completely implemented in Bluetooth 4.0 under the name Bluetooth Low Energy [BLU 15]. Launched in October 2011, the iPhone 4S was the first mobile phone to have this technology.

Regarding the technical characteristics, they are very similar to the classic Bluetooth. Nevertheless, the bit rate is a little lower, going up to around 200–300 kbps. Consumption, the Bluetooth LE's principal appeal, goes from 1 W to between 0.01 and 0.5 W [GAI 12].

3.3.4.2.3. ZigBee

Finally, ZigBee[13], based on IEEE standard 802.15.4, is a low-consumption communication protocol just like the majority of technologies intended for PAN. It aims to compete with Bluetooth by being simpler, less energy intensive, with a range that's more or less equal and less expensive.

ZigBee took its first steps in 1998, following the lead of Bluetooth and Wi-Fi. Its development was principally motivated by specific needs that competing technologies could not meet. Indeed, ZigBee is compatible with mesh networking. This organization reproduces the Internet functioning, each node can both receive and relay data. The information circulates from one node to another to the recipient node.

In 2005, the ZigBee Alliance[14] published the official specifications of the communication protocol. The data transmission distance extends from a dozen meters to a hundred for a bit rate set at 250 kbps [ZIG 16].

3.3.4.3. *Local and Wide Area Network (LAN/WAN)*

In addition to BAN and PAN wireless networks, there is also the LAN type of network at the higher level. This refers to a network of wireless communication operating in medium-sized spaces such as homes, offices, shops or museums. Most frequently, it is a domestic network made up of

---

13 The origin of the name ZigBee comes from the particular behavior that bees display to communicate remotely. Bees move in zigzags to send messages and provide directions to their fellow bees. This mode of communication is sometimes called the "Waggle Dance." Source: http://www.eetimes.com/document.asp?doc_id=1278172.

14 ZigBee Alliance is a group of manufacturers that maintain the ZigBee standard.

personal computers networking via a router for accessing the Internet. Wi-Fi is one of the wireless technologies that can integrate a LAN network.

The final category, WAN, as its name indicates, encompasses wireless communication technologies, whose range can cover a large geographical area. This category mainly includes the solutions used for cellphone networks.

3.3.4.3.1. Wi-Fi

Wi-Fi technology refers to a group of wireless communication protocols. It is integrated into almost all computers (desktop or portable) and is generally used to connect the computer to a router for Internet access:

> "Technology of wireless transmission by radio waves intended for a local network, which allows the exchange of data at a high bit rate and to have access to the Internet." (Office québécois de la langue française, http://grand dictionnaire.org, 2008)

Wi-Fi technology is based on the IEEE 802.11 standard. The latter was introduced in 1997 (date of the first version) after several experiments concerning wireless connectivity[15]. The term Wi-Fi was used for the first time for a commercial product in 1999 by the company Interbrand which is also the originator of the logo that resembles the yin-yang symbol.

"Wi-Fi" corresponds to a label issued by the Wi-Fi Alliance, the organization in charge of the devices interoperability specifications. Consequently, a device is marketed with the label "Wi-Fi" only if it is compatible with one of the 802.11 standards. Following the original standard (802.11), the IEEE developed new versions of the standard, each one of the versions bringing its lot of improvements and original specifications (see Table 3.3).

When it comes to applications, Wi-Fi is one of the preferred technologies for accessing the Internet. As previously emphasized, it is a technology which tends to be omnipresent, and which is implemented in the majority of mobile phones designed today. "Hotspots"[16] are multiplying in major cities. For example, the city of Paris has installed more than 300 Wi-Fi hotspots in public facilities [MAI 16].

---

15 In 1991, a technology named WaveLAN was developed for a cash register system.
16 A hotspot is a Wi-Fi terminal that allows access to the Internet.

Standard	Date	Maximum rate	Average range (indoors)	Average range (outdoors)
802.11	1997	2 Mbps	~20 m	~100 m
802.11a	1999	54 Mbps	~25 m	~75 m
802.11b	1999	11 Mbps	~35 m	~100 m
802.11g	2003	54 Mbps	~25 m	~75 m
802.11n	2009	450 Mbps	~50 m	~125 m
802.11ac	2014	1,300 Mbps	~20 m	~50 m

Table 3.3. *Some of the changes in the 802.11 standard [IEE 16c, IEE 16d, WIF 14]*

3.3.4.3.2. Cellphone networks

Cellphone networks have been developing quickly to respond to growing demand. Their infrastructure and technological solutions have been designed for a network used by millions of users simultaneously.

The telephone network is called cellular, since the territory covered by the network is divided into small areas called "cells". Each cell has a base station with a transmitter-receiver (an antenna) which is assigned a range of frequencies. The cells are hexagonal so that the distances between each adjacent cell is identical.

The base station of each cell is made up of an antenna, one or several transmitters-receivers and a controller. The latter manages the calls made by one of the terminals to the rest of the network. The base station is connected to a mobile switching center whose mission is the channel's assignment for calls, (to ensure the passage from the terminal of one cell to another) as well as recording of billing data.

The cellular network makes it possible to optimize the use of frequencies and thus guarantee the mobile phone services for all users. Among the cellphone networks, we can name the GSM, GPRS/EDGE, UMTS and LTE networks (see Table 3.4).

Networks	Description
GSM	Network at the base of the telephone communications. It was born along with mobile infrastructure and the network cellular mesh. Some inconveniences, it imposes different ranges of frequencies for adjacent cells to avoid the risk of interference and does not use packets for data transmission.
GPRS	Also called 2.5G, this type of network authorizes data transfers in packets with a theoretical maximum bit rate of 171 kbps. It corresponds to the Internet's first appearance on mobile telephones.
EDGE	The 2.75G network improves the speed of transmission of packets which is now 384 kbps.
UMTS	Better known under the name 3G, the bit rate was sufficient for sending emails, streaming video or photo sharing.
LTE	4G, the most recent generation. It allows bit rates able to theoretically reach 150 Mbps.
5G	The next generation of cellphone networks which should be introduced in 2020. It would allow bit rates of several gigabits per second and would be better adapted to Cloud Computing as well as to the IoT.

Table 3.4. *Summary of mobile cellular networks [GUP 13]*

Cellular networks are one of the possible solutions for ensuring communication between the different IoT actors: COs, gateways, data centers, data analysis platforms, online services, etc.

In light of the characteristics detailed in the table above (see Table 3.4), cellphone networks can potentially solve problems raised by certain constraints to which COs are subject. Indeed, these depend on the domain and the situation, but the CO installation can involve wireless connections over very long distances. In the area of connected agriculture, some applications required communications over kilometers, sometimes dozens, for example in the case of individual surveillance of each animal in a herd. Yet, cellphone networks properly cover vast geographical areas. The mobile phone operators Orange, Free and SFR have installed infrastructure for their networks extending all over France, which continue to grow.

In addition to the GSM, other networks offer attractive bit rates for COs. This is especially the case for GPRS and EDGE, both of which offer relatively low bit rates compared to LTE and UMTS nevertheless, they could be suitable for machine-to-machine communication. We saw this with the networks SigFox and LoRa, M2M is not very demanding in terms of bit rate and authorizes transmission speeds at this scale.

3.3.5. *Networking*

To sum it up, we have COs that are identifiable through unique addresses, that is to say IPv6 addresses. These objects are related to several sensors and actuators that, thanks to their multiple transducers, transpose the physical world which surrounds them into electrical signals. MEMS therefore serve as an interface between the physical world and electrical signals. Once these signals are generated, the data is exchanged between different devices for the purposes of obtaining a certain cohesion and cooperation in the actions carried out. This is Machine-to-Machine communication. Then, the signals produced by the devices are integrated into the virtual world thanks to communication technologies. They include, among others, Bluetooth LE, ZigBee, NFC or Wi-Fi technology. Finally, it is necessary to link this physical world made up of objects and this virtual world to platforms located in the Cloud and, to a lesser extent, to users.

Networking represents the last step necessary to enable the IoT. This step consists of connecting the IoT actors, that is to say to establish communication channels between COs and service providers. The data generated by the sensors must be transmitted to the service platforms. The businesses that produce COs set up, in parallel, platforms (based on the Cloud computing model) whose purpose is to process the data sent by COs. This process is at the heart of the services offered to users: using storage and data analysis, platforms can provide applications with more features. More useful and "smarter" applications as well as new functionalities, such as controlling a CO remotely, are born from data generation and analysis.

For this task, the Internet seems to be the best candidate because it already allows billions of internet users to communicate with each other[17]. In addition, transmitting information from one end of the Internet to the other is not possible without a protocol governing the transportation mode. Using the highway system analogy, people trying to get from point A to point B must set their itineraries, but also choose the transportation mode they will use for

---

17 The mark of 3 billion internet users was crossed at the end of the year 2014.

the trip. In the same way, data traveling through the Internet needs a transportation mode.

Protocols were introduced to standardize and govern communication means for objects through the Internet. In particular, CoAP, MQTT, AllJoyn and REST HTTP.

3.3.5.1. *Constrained Application Protocol (CoAP)*

COAP is an open standard adapted to electronic devices with limited resources (for example memory, energy, power) such as sensors. With this protocol, devices can communicate and interact through the Internet [ROS 15].

COAP was mostly edited by the IETF and more precisely by the task force called Constrained RESTful Environments (CoRE); the basic specifications are accessible in RFC 7252 [SHE 14]. The goal of this standard was to develop a communication protocol that was simpler than existing ones (HTTP, for example), but kept certain advanced specifications such as multicast[18]. During the conception, the CoRE task force is in charge of ensuring interoperability between CoAP and HTTP, and integrating support for the URI protocol. CoAP is compatible with all devices supporting UDP[19] [COL 11].

3.3.5.2. *Message Queuing Telemetry Transport (MQTT)*

MQTT is a communication protocol invented by Andy Stanford-Clark, an engineer at IBM, and Arlen Nipper of Cirrus Link Solutions in 1999.

MQTT is an ISO[20] standard dedicated to publish-subscribe communications and based on the IP/TCPTCP/IP protocols pair. This mode allows an emitter to transmit messages not directly to a recipient, but in a flow or a category. Thus, the recipients are those who have signed up for this flow or this category, they "subscribed" to it. The subscribers only receive messages that fall under the selected categories. Within this architecture, there is also what is called a message broker, that is to say a program in

---

18 Multicast is a technique aimed at sending packets from a single source to several sources.

19 User Datagram Protocol is a communication protocol used by the Internet. Like TCP, it oversees transporting packets, but unlike TCP there is no "handshaking" (negotiation) operation before each is sent.

20 ISO/IEC PRF 20922.

charge of receiving the messages emitted in order to transfer them in a format that is compatible with the recipients [BAN 15].

3.3.5.3. *AllJoyn*

Presented for the first time in 2011 at the Mobile World Congress by the company Qualcomm which is the principal initiator of the project, AllJoyn aims to be a system favoring the interoperability of COs and applications [ALL 16].

It is an open framework whose goal is to allow a device to communicate with other devices around it, independently of the brand. In other words, AllJoyn provides the necessary tools for ensuring communications between products of different origins and designs [LIO 11].

In addition, to go further with this open connectivity approach, the organization AllSeen Alliance was created to promote interoperability within the IoT.

3.3.5.4. *Representational State Transfer (REST)*

Developed in 2000 by Roy Fielding in his doctoral thesis, REST is an architecture designer for hypermedia systems which define the components, role and constraints involved during interactions between these same components. The architecture ensures that systems implement properties such as performance, simplicity, portability or flexibility.

Systems applying the REST architecture within the constraints related to interactions are called RESTful. RESTful systems usually work with the protocol HTTP and use verbs related to it: PUT, GET, POST and DELETE. These verbs correspond to the acronym CRUD that refers to the four core operations used to store information in the database which are respectively: CREATE, READ, UPDATE and DELETE. RESTful architectures are mostly implemented in web services[21].

Web services bring together a group of technologies which allow terminals (and their applications) to communicate between themselves, via Internet and independently of the languages they use. They rely on

---

21 While REST is more popular, some web services use the Simple Object Access Protocol (SOAP) whose implementation is more complex, with the benefit of better reliability than REST.

widespread protocols such as XML (Extensible Markup Language) or HTTP. These programs are accessible on the Internet by using web standards; therefore they have advantages that make them more and more popular, due to an increased mobility of users and the variety of devices used in everyday life. The interfaces of these programs are described in a way that is interpretable by machines, and client applications can access their services automatically. The use of languages and protocols independent of platforms reinforces the interoperability between web services.

3.4. Opportunities and threats in the IoT ecosystem

Originally the sole privilege of computers, today connectivity also includes physical objects, as they are becoming an integral part of the Internet network. The expansion of the Internet to "things" makes the border between the physical and virtual worlds more and more permeable. The quantification of the physical world by COs and their multiple sensors (for example temperature, movement, humidity, position, light) as well as the processing of measurements with sophisticated algorithms could profoundly transform entire sectors by changing the way we currently practice medicine or even by transforming entire industries.

This vision should be qualified: although applications and services developed around the IoT have the capacity to improve, optimize or even automate various activities, questions remain regarding the device's security, the confidentiality of data or the true added value of this data. There are dangers that accompany the IoT development and require the setup of regulation systems and independent monitoring organizations in order to avoid problems such as users abusive surveillance, the collection of information without authorization and the hacking of embedded systems.

3.4.1. *Opportunities*

Since 2010, the IoT has been moving in and growing in size, and businesses following this trend has developed strategies to integrate new products into various sectors such as medicine, industry, home automation or social networks. At the time of industry 4.0, French startups are active in the production of new COs, demonstrating dynamism, leading to a new generation of entrepreneurs, investors, engineers and designers gathered under the French Tech label. Besides, we see the appearance of various national or international organizations whose goal is to regulate the IoT development in one or several specific areas.

The IoT is the result of the Internet expansion not just limited to computers, but also extended to objects and people, a world "constantly connected" by means of a single and unique worldwide network where individuals share information and collaborate without technical limits. The unbridled online posting of personal information and the constant online exhibition of private life through social networks such as Facebook, Twitter or Instagram, these virtual public places, are disturbing the right to a private life that was previously the modern era prerogative. Based on this observation, in his book *La nouvelle société du coût marginal zéro*, Jeremy Rifkin suggests that the IoT goes hand in hand with the return to the public life that existed before the development of capitalism and is "[...] moving humanity from the era of private life [...] into the era of transparency" (p. 114).

Before the capitalist era, the right to a private life was nonexistent and everything, or almost everything, was done in public (sleeping, eating, grooming, etc.). Capitalism was accompanied by an increase in individualism, the appearance of the concept of property and the exclusion of "others": "the confinement and privatization of human life went hand in hand with enclosures – the confinement and privatization of the communal" (p. 115). Thus, private life was taken for a natural right by individuals and no longer as "[...] a social convention adapted to a particular moment in the human journey" (p. 115). Today this "natural" right collides with pervasive computing and the omnipresence of the Internet, individuals communicate and exchange between pairs displaying a desire for collaboration rather than exclusion. Again according to Jeremy Rifkin, we are in a "[...] an intermediate period between the capitalist era and the collaborative age [...]" (p. 116) and "[...] questions of private life will remain a major preoccupation, which will determine to a large extent the rapidity of the transition and the paths that we will take to enter into the next period in history" (p. 116).

For this reason, the European Commission issued a general principle to guide the development of the IoT [DIG 13]. Thus, the security of information and the protection of private life must be a part of the basic requirements for services related to the IoT. During the setting up of their applications, service providers are to deploy the necessary means to guarantee the security of the information and devices and preserve the confidentiality of the information they possess.

3.4.1.1. *Applications*

Among the potential applications, medicine is often held up as an example to demonstrate the extent of the possibilities offered by the IoT. Indeed, medecine will be profoundly transformed and become predictive and personalized. It will be personalized, because doctors will have the necessary means to individualize treatment for each patient. It will become predictive, thanks to the large volume of data generated by COs: Machine Learning algorithms will analyze this health data in real time and will alert the patients and doctors of any imminent risks and pathologies. The ongoing surveillance of vital signs will allow, for example, doctors to detect the flu before the patient begins to exhibit the first symptoms. By combining the curves of the heart rate, breathing and temperature, it is possible to identify the "V flu" phenomenon signaling the flu that is characterized by simultaneous lows in temperature and heart rate [CLA 16].

The smart house has been announced as the successor of home automation, not by replacing it, but rather by extending it. It is described as an environment where computer systems embedded in the different pieces and objects in the house can communicate with each other. The smart home would also be more economical thanks to functionalities that optimize energy consumption [GUB 13].

The automotive sector will also be transformed. Cars with drivers will give way to autonomous cars able to drive without human intervention. Close to 90% of automobile accidents are caused by human error [ASS 16], therefore driverless cars would make it possible to reduce the number of accidents[22] due notably to the fact that computer programs have a very short reaction time.

3.4.1.2. *The industrial Internet*

The industrial IoT or "Industrial Internet" is an IoT subcategory, a paradigm that suggests the transformation of industry as it exists today.

---

22 Nevertheless, algorithms also make judgment errors. On February 14, 2016, the first Google Car was responsible for a road accident. http://www.sciencesetavenir.fr/high-tech/drones/20160311.OBS6232/video-voici-le-film-de-l-accident-provoque-par-la-google-car.html.

Today's industry is a result of the third industrial revolution and is predominantly characterized by the development of sectors of activity based on information and communication technologies. Similarly, industry would undergo a fourth revolution by means of the more and more common use of cyber-physical systems and of the IoT (see Figure 3.11). To a certain extent, we are witnessing the digitization of industry [EVA 12].

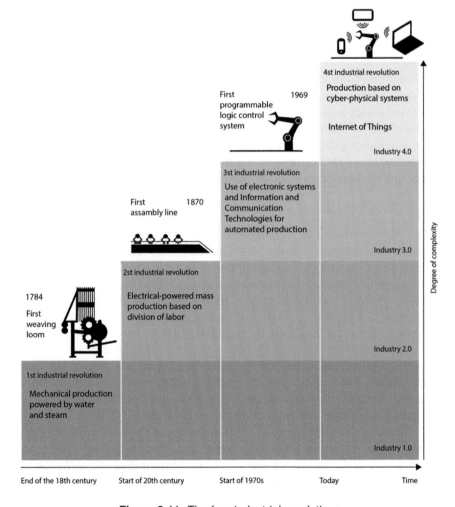

Figure 3.11. *The four industrial revolutions (adapted from DiePress.com, 2014 [DIE 14])*

Closely linked to robotics and networks of sensors, a cyberphysical system is a collaboration of computer systems in a network whose goal is to control and manipulate physical entities. This structure has characteristics similar to the IoT, namely networks of sensors and process automatization. The difference is that the connection of machines and sensors must satisfy critical manufacturing processes. In industries such as aeronautics, defense or aerospace, errors and crashes of devices have serious repercussions and can put human lives in danger. It is the same to a lesser extent for industries specializing in sports equipment or home automation. In sum, the level of requirements expected from computer systems and communication networks is higher [MIN 15].

Continuous communication of tools and computer systems provides significant advantages such as capacity for self-testing or remote control. The 4.0 industries will be more flexible and will make it possible to respond individually to consumer demands. In addition, manufacturers can carry out precise simulations thanks to the collection of data generated as well as the computerization of infrastructures. Finally, collected data can also be used to regulate and optimize energy consumption.

3.4.1.3. Governance

The development of a large variety of connected devices and unprecedented services raises new problems in regard to regulation, problems that have not existed until now. COs produce novel situations that are not subject to any regulatory body or even to the current legislation. The need for a judicial and technological framework is all the more important because the IoT growth is relatively rapid, according to the estimates that there will be several billion COs by 2020 [ROS 15].

Several international organizations were created around major companies such as IBM, Intel, Ericsson, Samsung or Microsoft to ensure these regulatory functions, such as the IoT Security Foundation (nonprofit organization dedicated to the security aspects of the IoT) [ERI 16] or the Open Connectivity Foundation (a proponent of better interoperability between devices originating with different manufacturers) [OPE 16]. Nevertheless, both multisectorial and multidisciplinary regulation is necessary in order to standardize the IoT development at the international level (the same way the Internet was regulated).

The regulation must obviously concern infrastructure, network architectures, COs and their interoperability as well as their safeguarding against hacking attacks, but also, and particularly, data collected by objects. Indeed, objects are designed to record all the information that surrounds them, such as the vocal assistants which are required "to listen" continually to the environment and user conversations in order to work. In a book published in 2015 [HOL 15], the authors Jonathan Holdowsky, Monika Mahto, Michael E. Raynor and Mark Cotteleer define four main principles related to data:

– knowledge and choice. The user must be notified in case of data collection and must be able to decide whether or not this data can be recorded;

– the purpose of the data collection and usage limitations. When they collect data, companies must notify the user and also explain the purpose of the operation. In addition, businesses commit to use the data only within a limited framework, with the knowledge of users;

– minimization. When they collect data, businesses only recover what is necessary for the service to function;

– responsibility and security. The data has a private character and constitutes an asset belonging to the user. As a result, businesses are responsible for the data collected and must make use of all the means necessary to guarantee data security (against data theft, for example).

Faced with the problems raised by data security and private life, the Swiss Federal Institute of Technology of Zurich is working on a project called "Nervousnet." In the beginning, it was a simple research platform dedicated to experiments by scientists using the IoT. The anonymous data collected makes it possible to analyze certain aspects of the society. Today, Nervousnet is a "digital nervous system" open and accessible to everyone, and capable of providing a detailed description of the world that surrounds us in a single place, while preserving individuals' private life [HEL 16]. The project falls within the collaborative governance approach, just like the development of the online encyclopedia *Wikipedia*.

3.4.2. *Threats*

Governance and creation of rules to restrict services and the use of data collected by COs is necessary in order to avoid problems. Although the IoT applications contribute to the advancement of sectors such as medicine or industry, there also remain concerns about the potential dangers of a "constantly connected" world.

According to Luc Ferry [FER 16], the "solutionist" optimism that drives ideologues of the collaborative economy and a large part of the transhumanism movement "has something Orwellian about it: this ideal of a society of universal connection and widespread transparency, this kindly totalitarian desire to control everything, to predict everything, this universe where everyone can know everything about everyone else, this open world where we will all be [...] continually listened to, scrutinized, decoded, this universe where our connected objects, from our scales to our refrigerators and even our watches, will continuously monitor our diets, the number of steps taken during the day, our heartbeats, our cholesterol levels and other fun things to make our lives totally normative. Welcome to Gattaca, a new era of human improvement and universal social control!"

3.4.2.1. Harmful uses

COs created by businesses dedicated to medical assistance, fitness or home management are aimed at improving the daily life of individuals. The idea is that COs via their "smart" functions complete tasks that would normally require human intervention. With this done, owners of COs have more time to spend on tasks they consider more pleasant (for example leisure, work, family life, etc.) [FLO 15].

The use of objects as assistants dedicated to unappealing and unappreciated tasks risks infantilizing individuals. Indeed, taking the example of a smartphone capable of managing a calendar and generating alerts, users have less need to think for themselves. Another example, individuals no longer have to worry about food with a smart fridge capable of managing fresh foods, creating shopping lists and which could possibly purchase the food itself.

According to the American researcher Evgeny Morozov[23], this infantilization would indirectly lead to a reduction in creativity and innovation in individuals and in our societies. COs being machines, they are not likely to commit errors and individuals continually assisted by these machines are not likely to commit them either. Yet, error is one of the

---

23 "The fact that a growing part of our lives is being modified by a technology based on sensors, and the fact that our friends and acquaintances can now follow us everywhere, are two innovations that will profoundly change the work of social engineers and legislators as well as many other benefactors." (E. Morozov, *Pour tout résoudre cliquez ici: l'aberration du solutionnisme technologique*, FYP Editions, 2014, p. 352).

factors which makes it possible to advance research, innovation and the human species more generally.

There also exists a field of the IoT dedicated to socialization, which promises exchanges between individuals and new kinds of objects. It turns out that COs can also have the opposite effect and end up isolating users. As with the mobile phone and Internet, individuals no longer need to leave their homes. This could be even more true with the IoT, since the goal of COs is to improve everyday life for individuals by freeing them from tasks involving travel such as shopping or socializing with friends.

Finally, more generally, these risks could lead to CO addictions, similar to smartphone addictions today. Individuals might no longer be able to live without using their connected devices.

3.4.2.2. Addiction to being "constantly connected"

Just like the mobile phone has become a vital tool for some people, the CO could also become a source of addiction. Psychological troubles could possibly appear following the excessive use of connected tools.

The fear of missing something (Fear of missing out) is a social anxiety that has developed along with information and communication technologies [PRZ 13]. Individuals who suffer from this problem are constantly afraid of missing an important event that could give them the opportunity to interact with others via social platforms (for example Facebook, Twitter and Snapchat). This fear is generally accompanied by a dependence on the Internet and connected tools. Afraid of missing any information, the affected people never disconnect.

Other problems can arise, such as the phantom vibration syndrome, a tactile hallucination that causes some people to have the impression that their mobile phone is vibrating even though it isn't [ROT 10]. The exact causes of these hallucinations are unknown, it is nevertheless possible that they are induced by excessive use of mobile phones.

The complete quantification of life is a potential source of risk for users. The monitoring of athletic activities is doubtless the most telling example, individuals can follow their performances in real time and in detail (for example the number of steps, heart rate, blood pressure, speed, etc.). The desire to measure every single activity can turn into an addiction.

3.4.2.3. *Security of data and equipment*

While COs are synonymous with progress in various domains, the fact remains that certain fears weigh on users [FLO 15]. Questions about the respect for private life and the responsible use of data collected by service providers are frequently in the news. The intrusive nature of COs as well as the tendency to "listen" to the surrounding environment without interruption and in real time are risk factors for the private life of users.

The collection of data from the physical world is the nature of the IoT and is a *sine qua non* condition for the proper functioning of the services offered by businesses. In addition, these services cannot work without certain personal information from consumers such as name and e-mail address. Users are required to agree to sharing their personal data with businesses, they nevertheless remain confidential.

In addition to the possible abusive practices by businesses vis-à-vis data, data vulnerability to hacking and data theft are problems that are just as important. The data centers where the users' personal data is stored can be subject to hacking. The data can fall into the hands of people whose intentions are dubious. For example, an insurance company could attempt to collect, possibly illegally, information concerning the health of its customers. Thus, the CO originally working toward the well being and health of the user could end up turning against him.

Cybercriminality is a menace for the IoT, since as a result of being connected to the Internet, the objects are accessible to hackers just like a computer would be. The FBI made an announcement encouraging owners of driverless cars to be vigilant, a fact that testifies to real risks. Several events have shown the vulnerability of connected devices against attacks, sometimes with serious consequences, with the potential to endanger lives. An autonomous car can be hacked and controlled remotely, a hacker could take control of a house and watch its inhabitants without their knowledge using surveillance cameras.

3.5. Conclusion

Still in its infancy, the IoT development is fragmented by businesses who are constructing proprietary models, to the detriment of interoperability between COs from other manufacturers. Like the Internet and the WWW, which rest on solid bases of proven technologies such as TCP/IP protocols, the HTTP communication standard as well as the URI unique naming system, the IoT actors do not currently have standard means to build a real ecosystem for the IoT. Nevertheless, the initiatives and technologies exist for setting up this new paradigm. We can distinguish five key steps: the unique identification of objects (for example IPv6, 6LoWPAN), data capture (for example MEMS, NEMS), connectivity (for example SigFox, LoRa), integration (for example Bluetooth LE, ZigBee, NFC, RFID, Wi-Fi) and networking of COs (for example CoAP, MQTT, REST) with the Internet. In addition to the technical aspects that make up for many of the challenges for the future, the IoT raises ethical and legal questions. The growth of the paradigm goes hand-in-hand with the creation of a legal and technological framework as well as independent international organizations. Just like the Internet did, the IoT is ready to profoundly transform large parts of our society.

3.6. Bibliography

[ALL 16] ALLSEEN ALLIANCE, available at: https://allseenalliance.org/, 2016.

[ASA 14] ASARE D.A.K., Body Area Network – Standardization, Analysis and Application, Thesis, Savonia University of Applied Sciences, 2014.

[ASS 16] ASSOCIATION PRÉVENTION ROUTIÈRE, "Causes accidents de la route", available at: https://www.preventionroutiere.asso.fr/2016/04/22/statistiques-daccidents/, 2016.

[BAN 15] BANKS A., GUPTA R., MQTT Version 3.1.1, OASIS Standard, 2015.

[BLU 15] BLUETOOTH SIG, "Press Releases Details" *Bluetooth.*, available at: https://web.archive.org/web/20150203053330/, http://www.bluetooth.com/Pages/Press-Releases-Detail.aspx?ItemID=138, 2015.

[BLU 16] BLUETOOTH SIG, Bluetooth Technology Website, available at: https://www.bluetooth.com/, 2016.

[CIS 16] CISCO, "Internet of Things (IoT)", *Cisco*, available at: http://www.cisco.com/c/en/us/solutions/Internet-of-things/overview.html, 2016.

[CIV 12] CIVERA D., "Le world des MEMS RF – MEMS: le monde microscopique de votre smartphone", *Tom's Hardware*, available at: http://www.tomshardware.fr/articles/MEMS,2-811-12.html, 2012.

[CLA 16] CLAVERIE H., DEVOS N., MESSIKA S., "IoT et Big Data au service du Machine Learning", *ROOMn*, Monaco, 2016.

[CNR 16] CNRFID, "Introduction à la RFID", *Centre National RFID,* available at: http://www.centrenational-rfid.com/introduction-a-la-rfid-article-15-fr-ruid-17. html, 2016.

[COL 11] COLITTI W., STEENHAUT K., DE CARO N., "Integrating Wireless Sensor Networks with the Web", *Extending the Internet to Low power and Lossy Networks*, Chicago, 2011.

[CUR 12] CURRAN K., MILLAR A., MC GARVEY C., "Near Field Communication", *International Journal of Electrical and Computer Engineering (IJECE)*, vol. 2, no. 3, pp. 371–382, 2012.

[DIE 14] DIE PRESSE, "Industrie 4.0: Wenn die Revolution nach Österreich kommt", available at: http://diepresse.com/home/alpbach/385 8672/Industrie-40_Wenn-die-Revolution-nach-Osterreich-kommt, 2014.

[DIG 13] DIGITAL AGENDA FOR EUROPE, IoT Privacy, Data Protection, Information Security, A Europe 2020 Initiative, 2013.

[ERI 16] ERICSON S., Home, IoT Security Foundation, available at: https://iot securityfoundation.org/, 2016.

[EVA 12] EVANS P.C., ANNUNZIATA M., Industrial Internet: Pushing the Boundaries of Minds and Machines, Report, Imagination at work, 2012.

[FER 16] FERRY L., *La révolution transhumaniste*, Plon, Paris, 2016.

[FLO 15] FLORENT E., MANCEAU M., RAMAGE M. *et al.*, Approche sociologique des connected objects, Master's dissertation, Université Aix Marseille, 2015.

[GAI 12] GAIA ZANCHI M., *Bluetooth Low Energy*, LitePoint, 2012.

[GIL 12] GILMER B., "Star Wars Episode 1: The Phantom Menace Electronic CommTech Reade", available at: https://www.youtube.com/watch?v=5stDGP0 e05A, January 2012.

[GRA 06] GRABIANOWSKI E., "Is Wibree Going to Rival Bluetooth?", *HowStuffWorks*, available at: http://electronics.howstuffworks.com/wibree.htm, 2006.

[GRE 14] GREEN J., "Building the Internet of Things. An IoT Reference Model", *2014 Internet of Things World Forum*, Chicago, United States, 2014.

[GUB 13] GUBBI J., BUYYA R., MARUSIC S. *et al.*, "Internet of Things (IoT): A Vision, Architectural Elements, and Future Directions", *Future Generation Computer Systems*, vol. 29, no. 7, pp. 1645–1660, 2013.

[GUP 13] GUPTA P., "Evolvement of Mobile Generations: 1G to 5G", *International Journal for Technological Research in Engineering*, vol. 1, pp. 152–157, 2013.

[HEL 16] HELBING D., *FuturICT Blog: NERVOUSNET – Towards an Open and participatory, Distributed Big Data Paradigm*, FuturICT Blog, available at: http://futurict.blogspot.com/2016/01/nervousnet-towards-open-and.html, 2016.

[HÖL 14] HÖLLER J., *From Machine-to-Machine to the Internet of Things: Introduction to a New Age of Intelligence*, Elsevier Academic Press, Amsterdam, 2014.

[HOL 15] HOLDOWSKY J., MAHTO M., RAYNOR M.E. *et al.*, *Inside the Internet of Things (IoT)*, Deloitte University Press, Westlake, 2015.

[IEE 07] IEEE SPECTRUM, "Oops! How Many IP Addresses?" *IEEE Spectrum*, available at: http://spectrum.ieee.org/tech-talk/semiconductors/devices/oops_how_many_ip_addresses, 2007.

[IEE 16a] IEEE STANDARDS ASSOCIATION, "IEEE SA – P2413 – Standard for an Architectural Framework for the Internet of Things (IoT)", available at: https://standards.ieee.org/develop/project/2413.html, 2016.

[IEE 16b] IEEE P2413 WG, "P2413 Working Group", *IEEE P2413 Working Group*, available at: http://www.ieee802.org/11/Reports/802.11_Timelines.htm, 2016.

[IEE 16c] IEEE STANDARDS ASSOCIATION, "IEEE-SA-IEEE Get 802 Program – 802.11: Wireless LANs", *IEEE Standards Association*, available at: http://standards.ieee.org/about/get/802/802.11.html, 2016.

[IEE 16d] IEEE 802.11 WG, IEEE 802.11, "The Working Group Setting the Standards for Wireless LANs", *IEEE 802.11 Working Group*, available at: http://www.ieee 802.org/11/Reports/802.11_Timelines.htm, 2016.

[INT 16] INTEL, A Guide to the Internet of Things Infographic, Intel., available at: http://www.intel.com/content/www/us/en/Internet-of-things/infographics/guide-to-iot.html, 2016.

[KAR 16] KARLSSON S., LUGN A., "The History of Bluetooth - Ericsson History", available at: http://www.ericssonhistory.com/changing-the-world/Anecdotes/The-history-of-Bluetooth-/, 2016.

[KWA 10] KWAK K.S., ULLAH S., ULLAH N., "An Overview of IEEE 802.15.6 Standard", *International Journal of Engineering Research and Applications*, vol. 5, no. 12, pp. 1–6, 2010.

[LIO 11] LIOY M., "Peer-to-Peer Technology: Driving Innovative User Experiences in Mobile", *Qualcomm Innovation Center*, 2011.

[MAI 16] MAIRIE DE PARIS, "Open Data Paris – Liste des sites des hotspots Paris Wi-Fi", available at: http://opendata.paris.fr, 2016.

[MIN 15] MINERVA R., BIRU A., ROTONDI D., "Towards a Definition of the Internet of Things (IoT)", *IEEE Internet Initiative*, 2015.

[MOR 14] MOROZOV E., *Pour tout résoudre, cliquez ici: l'aberration du solutionnisme technologique*, FYP, Limoges, 2014.

[MUL 07] MULLIGAN G., "The 6LoWPAN architecture", *EmNets '07 Proceedings of the 4th Workshop on Embedded Networked Sensors*, pp. 78–82, Cork, Ireland, 2007.

[OFF 08] OFFICE QUÉBÉCOIS DE LA LANGUE FRANÇAISE, "Identification par radiofréquence – Le grand dictionnaire terminologique", available at: http://granddictionnaire.com/ficheOqlf.aspx?Id_Fiche=8362543, 2008.

[OPE 16] OPEN CONNECTIVITY FOUNDATION, "Open Connectivity Foundation (COF)", available at: http://openconnectivity.org/, 2016.

[PRZ 13] PRZYBYLSKI A.K., MURAYAMA K., DEHAAN C.R. *et al.*, "Motivational, Emotional, and Behavioral Correlates of Fear of Missing Out", *Computers in Human Behavior*, vol. 29, no. 4, pp. 1841–1848, 2013.

[POO 15] POOLE I., "LoRa Wireless for M2M & IoT", *RadioElectronics.com.*, available at: http://www.radio-electronics.com/info/wireless/lora/basics-tutorial.php, 2015.

[POS 81] POSTEL J., Internet Protocol, Request For Comments, DARPA Internet Program, 1981.

[POS 16] POSTSCAPES, "Internet of Things Technologies", *Postscapes*, available at: http://postscapes.com/Internet-of-things-technologies, 2016.

[ROT 10] ROTHBERG M.B., ARORA A., HERMANN J. *et al.*, "Phantom Vibration Syndrome Among Medical Staff: a Cross Sectional Survey", *BMJ*, vol. 341, p. 6914, 2010.

[RIF 16] RIFKIN J., CHEMLA F., CHEMLA P., *La nouvelle société du coût marginal zéro: l'Internet des objets, l'émergence des communaux collaboratifse et l'éclipse du capitalisme*, Babel, Paris, 2016.

[ROS 15] ROSE K., ELDRIDGE S., CHAPIN L., The Internet of things: An Overview, The Internet Society (ISCO), 2015.

[SHE 14] SHELBY Z., HARTKE K., BORMANN C., The Constrained Application Protocol (CoAP), Request For Comments, Internet Engineering Task Force, 2014.

[SIM 12] SIMONDON G., *Du mode d'existence des objets techniques*, Editions Aubier, Paris, 2012.

[TSC 15] TSCHOFENIG H., ARKKO J., THALER D. *et al.*, Architectural Considerations in Smart Object Networking, Request For Comments, Internet Architecture Board, 2015.

[WAT 14] WATRIGANT T., "Sigfox: comprendre la technologie M2M du fleuron français de l'Internet des objets", *Aruco.*, available at: https://www.aruco.com/2014/09/sigfox-m2m/, 2014.

[WG4 16] WG42, "Survey of Architecture Frameworks", *ISO-Architecture.org.*, available at: http://www.iso-architecture.org/ieee-1471/afs/frameworks-table. html, 2016.

[WIF 14] WI-FI, "15 Years of Wi-Fi | Wi-Fi Alliance", *Wi-Fi.*, available at: http://www.wi-fi.org/discover-wi-fi/15-years-of-wi-fi, 2014.

[ZIG 16] ZIGBEE ALLIANCE, "The ZigBee Alliance. Control Your World", *ZigBee.org.*, available at: http://www.zigbee.org/, 2016.

Toward a Methodology of IoT-a: Embedded Agents for the Internet of Things

4.1. Introduction

With the increase in the number of connected objects around us, the concepts of the IoT (*Internet of Things*) and Big Data are expanding to cover a wide variety of areas of application. The term "IoT" has been adopted successfully by the world of industry. However, less publicized research on the concepts of ambient intelligence, diffuse intelligence or smart objects has been going on for years. When it comes to this term, there are still many hurdles that need to be overcome. Much of the research done on the IoT is based on its architecture, the control of connected objects [KIR 15], their reasoning, sensors and effectors or the resources available to them [MAM 12, DUJ 11]. Different works [COU 12, SER 93] underline the parallels that can be made between research on multi-agent systems and the field of the IoT, for example, parallels regarding interactions, communication protocols, interoperability and autonomous behavior.

Our main problem is to conceive and experiment with an embedded multi-agent platform to allow different connected objects to interact autonomously. Our approach is based on the concept of *spime*, a neologism introduced by Sterling [STE 05]. A *spime* is an object localized in space and time and strongly associated with its history and the data that it carries within itself. The difference between the material and immaterial in the data

Chapter written by Valérie RENAULT and Florent CARLIER.

is reduced to its minimum possible. In this chapter, we view the objects of the IoT as the agents in multi-agent systems, with the goal of modeling and implementing a multi-agent architecture in the field of the IoT. Our connected objects thus become *Internet of Things – agents*, which we will call IoT-a. In this way each object integrates, in its conception, communication protocols implemented as closely as possible to electronic components.

Our research deals with a low-level embedded multi-agent architecture, called Triskell3S, and, in particular with the underlying communication protocols. In accordance with the FIPA-ACL standard, each agent is registered with the platform and is thus recognized by the other agents. Triskell3S is based on the interaction approach developed in IODA [KUB 11, MAT 13]. This model, initially created for the field of simulation, is here adapted into a real physical environment for the IoT-a.

In this chapter, we will have a quick review of the different paradigms of the IoT and the links that have been established in the literature between the IoT and multi-agent systems. We then present the embedded multi-agent platform Triskell3S by showing how the different paradigms and norms of the two domains can be met and coexist, in particular the MQTT protocol [LAM 12b], the D-Bus Protocol [DBU 16] and the FIPA-ACL standards. In order to experiment our architecture in a real-world context, we present an application of IoT-a through a set of connected "brick-screens" that allows us to build an interactive and reconfigurable wall of screens. We illustrate this application by revisiting the eco-distributed resolution of the N-Puzzle algorithm and applying it to the resolution of an N-puzzle video.

4.2. Multi-agent simulations, ambient intelligence and the Internet of Things

The growing complexity and new dynamics of information systems requires relying on new models and adaptive computer architectures. Numerous works on multi-agent systems (MAS) have demonstrated their capacity for proactivity, adaptation and auto-organization, allowing the molding of heterogeneous, dynamic and responsive systems to the needs of growing interactions with users [GDR 13, GUE 12, SAB 14]. Multi-agent systems have often been tested within the framework of mobile environments, in the context of ambient intelligence or the IoT. Platforms specific to these contexts such as JADE [BEL 17] or SPADE [GRE 06] have been developed.

The "agent" themes have evolved in two domains of research: artificial intelligence and distributed programming [CAO 12]. The goal of the latter works is to allow complex tasks to be executed with cooperation mechanisms between different agents distributed over a network. It will therefore be interesting to analyze these works in the light of connected objects, heterogeneous and distributed in the home environment. With the rise of the Internet of Things, access to information and digital services is "spreading" into living spaces and is more and more natural and user-friendly [GLE 14].

Numerous works have been developed in the area of ambient intelligence [COU 08], in particular on the contribution of agents to living areas with "intelligent" homes [ABR 09, CHO 12]. Many of these works have the goal of optimizing energy consumption in a building according to the habits of its occupants [MAM 12], of the user's comfort [CAR 06, MOZ 05] or the improvement of services and personal security [BRD 09, SUB 14]. Smartgrids, or smart energy networks, offer new infrastructures in order to improve the reactivity and reliability of networks according to supply and demand. Ramchurn's article [RAM 11] is a good summary of the state of the art done in this context with the integration of multi-agent systems. Here, the interest in agents is to learn from user habits by establishing profiles. For example, Mamadi *et al.* [MAM 12] present a network of adaptive agents dedicated to optimizing the home heating and ventilation. The knowledge the agents have of their environment is based on a set of sensors allowing the detection of movement, sounds, CO_2 levels, etc. These sensors have been developed specifically for this experiment. *Sensing agents* then collect this information so it can be processed by *prediction agents*. In the same context, with SAVES [KWA 12], the authors present an application that uses two types of agent. *Room agents* gather information from sensors present in the simulation and in the real environment. Then the *proxy agents*, available on a mobile terminal, allow the user to define his preferences in order to establish models of behaviors. These works, although they have electronic boards for the sensors, do not use these cards to directly carry the agents. As in the literature and the last two articles, the sensors are decorrelated from the agents and there are no specific studies on the low-level integration of the agents into the embedded board.

Different models and *framework agents* have also been developed in the context of the smart home, whether it is with BDI agents [SUB 14] or with more reactive agents. Prediction models of human activities are often at the base of these ambient systems whether they are models with a more or less cognitive base [DAV 98, MOZ 05, SAB 14] or experiments carried out with

Markov's models [RAO 04]. These works are based on the prediction of future actions using the recognition of sequences of past actions. This problem is not easy to solve in a living space since different people can interact with the system and give opposite orders. It is therefore appropriate to offer users of the home a coherent collection of multimedia data through different interactive devices. The works presented above regarding ambient intelligence very often use very simple embedded boards to support a set of sensors to recover information from the environment. Nevertheless, cards are just used to transmit information. For example, JADE facilitates the development of interoperable multi-agent systems. It "allows developers to implement and deploy multi-agent systems, including agents running on wireless networks or material devices that have limited resources" [BER 14]. In most cases, a *middleware* layer is provided to go between the hardware layer and the multi-agent application level. The multi-agent level is often based on simulation models and does not always take the underlying hardware constraints into account. Typically, these platforms are based on the Java language, a programming environment that is not optimal in an embedded context.

Research has been conducted in order to integrate agent algorithms into a graphics processing unit (GPU), especially in the domain of video games or simulations involving large sets of individuals. This is the case in the work of Da Silva [DAS 10] which compares two implementations of different GPUs in order to parallelize Reynold's Boids algorithm. Research on the generation of a correlation structure of coalition of agents with a GPU [PAW 14] presents an interesting approach that is close to the performance criteria used in research on GPUs.

This article underlines three important points, which must also be considered in the developed agent architecture:

– minimize the number of synchronization points;

– minimize the overall number of memory accesses;

– maximize the number of processes running in parallel.

When it comes to connected objects, middleware architectures are gaining more and more importance in the field of the IoT, with the adoption of the principles of Service Oriented Architecture (SOA) illustrated in Figure 4.1 [ATZ 10]. In these architectures, the middleware is based on five layers: at the higher level is the application layer dedicated to a given problem; at the lower level are physical objects that interact with their environment.

Between these two layers are layers of composition, management and abstraction that allow the gradual providing of services at a high level of abstraction to the final user.

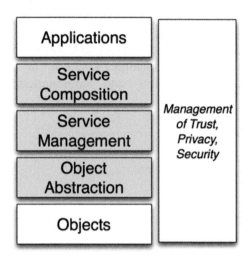

Figure 4.1. *SOA architecture for the IoT*

Multi-agent platforms are generally seen as the applicative layer of the SOA. This layer is therefore not embedded in an electronic component. In the majority of this research, "intelligence" is delocalized and calculated on the centralized server. The environment is made up of sensors and effectors without any real intelligence and interaction with each other.

Our works are close to those of Jamont [JAM 10, JAM 14] on the DIAMOND (*Decentralized Iterative Multiagent Open Networks Design*) method. He presents a method of co-design specialized in the realization of an embedded system integrating an MAS. Thus, DIAMOND concentrates on the realization of an embedded system adapted to the physical needs of a multi-agent system. As far as we are concerned, the goal is to encapsulate the architecture agents closest to the hardware components.

4.3. Triskell3S: an architecture of embedded agent-oriented inter-actions

Triskell3S is an embedded architecture based on an agent-oriented approach to interactions. The acronym "3S" stands for Embedded Systems, Multi-agent Systems and Mobile Systems. This architecture respects the definition of the IoT provided by the European Commission (2008): "Things having identities and virtual personalities operating in smart spaces using intelligent interfaces to connect and communicate within social, environmental, and user contexts."

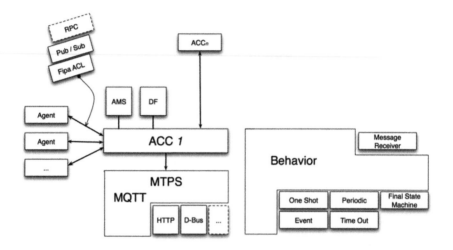

Figure 4.2. *The embedded multi-agent architecture of the Triskell3S platform*

Figure 4.2 presents the architecture of the Triskell3S platform. The platform is based on the FIPA (*Foundation for Intelligent Physical Agents*) standard. It offers the implementation of agents to ensure administrative control of the whole multi-agent system (ACC for agent communications channel, AMS for agent management system, and DF for service book). For transferring messages between agents, we use the ACL (*Agent Communication Language*) standard. This standard provides a rule for encapsulating messages. The MTPS (*Message Transport Protocol Service*) makes it possible to avoid the physical means of data transfer (HTTP, D-Bus, MQTT, etc.).

The platform is based on two levels of inter-agent and intra-agent interactions. The first level implements the MQTT protocol, at the heart of the platform [LAM 12], while still meeting the FIPA-ACL communication specifications [CON 03] that allow the IoT-a to interact. The MQTT protocol is based on the Publisher/Subscriber mechanism. The client subscribes to a topic via the MQTT platform known as BROKER. Multiple clients can subscribe to a *topic*. They will then all be listening to all the posters writing about the *topic*. The BROKER and the MQTT have simple speech acts that are especially adapted to the IoT. At the second level, inter-process communication within the agent itself is based on the D-Bus Protocol (message system using sockets) [PAT 04]. This protocol allows us to shift from the software layer to the hardware one. Each agent is autonomous and functions via the use of a *thread* with which we can associate behaviors to personalize its characteristics for the overall resolution of a problem. On our Triskell3S platform, we can find the three important points underlined by Pawlowski's article [PAW 14] cited previously.

4.4. Transposition of the formalization of agent-oriented interaction to connected objects

In order to illustrate our architecture in a real-world context, we have defined a heterogeneous set of connected objects that allow the composition of a wall of synchronized and interactive screens. The goal is to show how agent-oriented approach interactions, introduced within the electronic boards of each object themselves, can make it possible to build a coherent system.

If we look purely at the application, research on screen walls (or walls of images) have existed for many years [BEA 12, NAN 15, PIE 11] in numerous contexts (television, CCTV, etc.), including curved or touch screens. Numerous integrated solutions with large visualization supports have existed since the 2000s.

Generally, there are three types of architecture:

– different screens are linked to the same office on a single computer (a monitoring system);

– each screen is connected to a specific monitoring system, for example with cameras (CCTV, detection and conflict prevention), in order to display different content on a multitude of monitors;

– the same video is displayed on a group of screens; this generally requires pre-processing in order to adapt the portion of the image to be displayed on each screen, depending on the number of screens present.

Our images wall is a support for testing the architecture of IoT-a and demonstrating the effectiveness of their communication protocols. Indeed, in the context of a wall of images, the synchronization of different screens in time becomes a real challenge, especially with autonomous "bricks" consisting of screens. If we want our system to adapt in real time, each screen must be capable of informing the other screens in the zone displayed. It is therefore necessary for it to communicate and interact with the other display modules. Figure 4.3 summarizes the organization of the IoT-a in this context. To do this, each screen is paired with an embedded board in order to form a programmable brick for visualization. In our experiment, we have chosen to pair each screen with a Raspberry Pi [UPT 12] embedded board. These boards have the advantage of being light (the size of a credit card) and including a graphics processor powerful enough to assure the fluidity of the videos. An autonomous brick corresponds to an IoT-a. The interactions between each IoT-a provide us with different types of display. Different videos can be displayed independently on each of the screens. A single synchronized video can be positioned over the entire wall of images. It is also possible to have N videos synchronized on a predefined number of screens M.

Raspberry Pi boards have already been used to build walls of images, such as in the works around the PiWall Project [GOO 15]. These works are based on a master/slave architecture that is made up of a card per screen and a supplementary master card that makes it possible to control and synchronize the whole. In Triskell3S, this latter master card disappears. All of the bricks are identical in terms of architecture and role but also in terms of awareness of their environment. The coherence of the set of IoT-a rests on the communication protocols and interactions.

In the field of agents, the works of Rihawi [RIH 13] present a study on the impact of different synchronization policies (strong synchronization, synchronization with a window of time and absence of synchronization) at the macroscopic level of multi-agent simulations. Nevertheless, these works have been tested within multi-agent simulations. In terms of the screens wall the goal is to embed agent models in real environments, which adds new constraints for synchronization, performance, the cost of inter-object communication security and stability of the system. Optimizing the interactivity in the walls of images is a necessary challenge for supporting collaborative applications [CHA 14]. For our experiment, and in order use a heterogeneous system, we have introduced specific IoT-a in order to interact on the videos, in the same way as the KinectAgent, QRCodeAgent and SMSAgent.

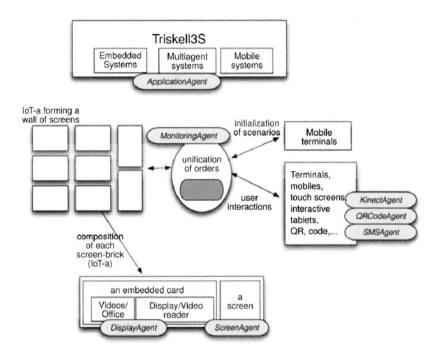

Figure 4.3. *Implementation of Triskell3S on a set of IoT-a collectively forming a screens wall*

As their name suggests, these three agents are connected to specific hardware components, allowing the user to send commands to the wall. For example, the command STOP send by the *SMSAgent* makes it possible to stop all the videos at once or only a few videos if this command is sent to only certain IoT-a. The MOVE command makes it possible to move a video from one screen to another. Other classic commands have been specified such as PAUSE, MUTE_VIDEO, etc. This architecture therefore allow us to deploy a set of specific components dedicated to the control of and interaction with the videos. Nevertheless, on the user side, the method of interacting with the data displayed according to the usage context still remains to be developed.

The tree of MQTT topics (Figure 4.4) shows the topic hierarchy used in the Triskell3S platform. We define the ACC topic representing the communication channel agent. All the agents are identified and listening to their identifier in this channel in order to be able to interact together (ACC/agent-id.name/). Then, we find the channels for discussion with AMS and DF which serve to register the agents and their services. In the N-puzzle experiment, an agent organization called OMX groups together all of the agents managing the screens. The agents react to the *topic* ALL for a global command or to their unique identifier OMXid. Two classes of command have been defined for interacting with the agent screens: CMD and CMDX. The *topic* CMDX corresponds to the textual command, for example WALL_ACTIVE or MUTE_VIDEO. CMD are equivalent but with the command identifiers WALL_ACTIVE: 15 and MUTE_VIDEO: 39. We therefore have the possibility of adapting our ontology with the command (CMD).

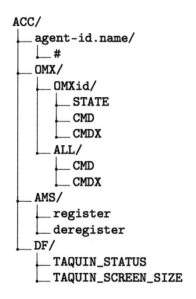

Figure 4.4. *Directed tree of MQTT Topics*

4.5. Formalization

We have tested our architecture on a classic model in the field of multi-agent systems by applying the eco-resolution N-puzzle problem [DRO 93] to our wall of images. Traditionally, the N-puzzle problem is based on a square board containing N tiles and an empty position known as the blank position. The goal is to move the tiles arranged in a random position toward a final configuration, making it possible to reconstruct a coherent global image. This problem has been formalized in numerous works by tile agents, each agent being characterized by a state, a goal, a behavior of satisfaction and flight of other agents.

In our experiment, and in order to ellustrate the Triskell3S platform on the wall of IoT-a, the N-puzzle problem becomes the distributed resolution of a N-puzzle video which has the goal of reconstructing a final synchronized video. The goal of this chapter is not to offer a new formalization of the N-puzzle, but rather to experiment our architecture of embedded agents against an already-known decentralized problem and in this way to test the effectiveness of our communication protocols. To do this, we will first show how the formalism of the N-puzzle can adapt to the IoT-a in a more general way.

In the distributed approach of the N-puzzle [DRO 93], the overall G_{puzzle} goal is broken down into n smaller independent goals that must be met so that the G_{puzzle} is satisfied. The subgoals representing the position of each tile at its final location is described by:

$$G_{puzzle} = \{position(\tau_1, p_1), \ldots, position(\tau_n, p_n)\} \qquad [4.1]$$

where n is the number of tiles, τ_i represents the tiles and p_i the final states to reach (the goal state).

In our case, a tile represents a portion of the video and state corresponds to a physical screen positioned on the wall. It is also necessary to consider the display software controlling the video on the screen. The device is therefore defined by:

$$G_{puzzle} = \{position(v_1, s_1, d_1), \ldots, position(v_i, s_n, d_j)\} \qquad [4.2]$$

where v_i, s_n, d_j represent the video on each tile, the physical screen and the *display* (the software that plays the video) respectively. The index i, n and j allow us to have different displays and thus different videos on the same screen. A screen can then support several tiles at the same time if you would like to use a board larger than the number of screens available.

The distinction between the physical screen and the *display* of the rendering is important in the context of an embedded system and, therefore, in our architecture. It directly effects the management of messages between the agents. If we consider this formalism as a more general context for the definition of a *framework* for IoT-a, the video is the "data" used by the agent, the screen is the hardware component that supports the agent and the *display* makes it possible to link between the data and the hardware components. We can therefore describe the subgoal of an IoT-a in the following matter:

$$G_{env} = \{handler(d_1, h_1, c_1), \ldots, handler(d_i, h_n, c_j)\} \qquad [4.3]$$

where d_i, h_n, c_j represent respectively the data to be processed, the hardware component and the software layer of communication.

Here also, indexes can be different. Thus, the amount of data (videos read at a precise moment) is not tied to the number of hardware components.

We change point of view and view the N-puzzle in terms of agents, by considering an agent, a_i, with its state, *state(a_i)*, and its *behavior(a_i)*. Behavior represents all of the actions to be carried out in order to attain a goal from a current state. So, the current state is defined by the authors [DRO 93] as:

$$\forall g_i \forall a_i / goal(a_i) = g_i, satisfied(g_i) \Leftrightarrow state(a_i) = goal(a_i) \quad [4.4]$$

Next, the authors [DRO 93] apply this definition to the N-puzzle, which means we can define a tile τ_i as:

$$\tau_i = \{p_i, p_k, behavior(\tau_i)\} \qquad [4.5]$$

$$satisfied(position(\tau_i, p_i)) \Leftrightarrow p_k = p_i \qquad [4.6]$$

where p_i is the final target position of the tile τ_i, p_k is the current position and *behavior(τ_i)* is the set of actions to carry out in order to move the tile from its current position to its final position.

With formulas [4.5] and [4.6], the authors show how the satisfaction of a sub-goal rests on the agent tile's capacity to "do the right thing" to attain its goal and in this way to offer a description of problem-oriented agents.

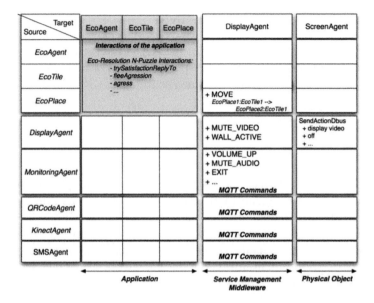

Target Source	EcoAgent	EcoTile	EcoPlace	DisplayAgent	ScreenAgent
EcoAgent	*Interactions of the application*				
EcoTile	*Eco-Resolution N-Puzzle Interactions:* *- trySatisfactionReplyTo* *- fleeAgression* *- agress* *- ...*				
EcoPlace				+ MOVE *EcoPlace1:EcoTile1 --> EcoPlace2:EcoTile1*	
DisplayAgent				+ MUTE_VIDEO + WALL_ACTIVE	SendActionDbus + display video + off + ...
MonitoringAgent				+ VOLUME_UP + MUTE_AUDIO + EXIT + ... **MQTT Commands**	
QRCodeAgent				**MQTT Commands**	
KinectAgent				**MQTT Commands**	
SMSAgent				**MQTT Commands**	

◄──────────► ◄──────────► ◄──────────►

Application   *Service Management*   *Physical Object*
 Middleware

Figure 4.5. *Matrix of IODA interactions in the context of IoT-a making up a wall of screens. An example of the distributed resolution of the N-puzzle problem*

However, in the context of the IoT, [4.3] shows the importance of communication and interaction between agents. This allows us to switch from an agent-oriented approach description to an interaction-oriented approach, as defined in the IODA methodology (*Interaction-Oriented Design of Agent Simulations*) [KUB 08]. The concept of interaction is very close to the concept of *affordance* defined by Norman [NOR 88]. The idea is that the objects carry within themselves the knowledge of their use. IODA methodology consists of separating the actions in which they participate in order to reify them into the concept of interaction. The authors propose defining a matrix of interactions between source agents and target agents. Each cell of the matrix corresponds to the actions that a source agent can carry out on a set of target agents. Thus, in the context of embedded systems, we propose adapting this model, initially dedicated to simulations, by developing two levels of interaction. The first level consists of defining inter-agent interactions based on the MQTT communication protocol. Then we define the level of intra-agent interaction allowing communication between the hardware layer and the software layer. The inter-process communication within the agent itself is then based on the D-Bus Protocol. These different layers of communication remain compatible with the organizational level of the SOA defined in Figure 4.1.

4.6. Experimentation and perspectives

Figure 4.5 presents an adaptation of the IODA interaction matrix applied to the distributed resolution of the N-puzzle by IoT-a and illustrated by a wall of images. This adaptation is defined as follows:

$$+(interaction, range, priority) \qquad\qquad [4.7]$$

$$+(interaction) \qquad\qquad [4.8]$$

The IODA model defines the range and priority in the definition of the interaction between two agents [4.7]. These two conditions are relevant within the framework of simulations of social behaviors or within a physical framework of interaction. In an embedded context, the concept of range is hidden by the interaction via the connection to the hardware data bus (D-Bus). The concept of priority is useful within the framework of parallel actions that require a starting point or a decision to be hardware. The rest of the time, the interactions themselves follow a logical order of resolution. In terms of the resolution of the N-puzzle, we have chosen to simplify the writing of our interactions by not giving priority in the case of interactions [4.8].

Figure 4.6. *Distributed resolution of the N-puzzle on the wall of agentified screens. Each portion of the video represents a piece of data carried by a tile. The goal is to reconstruct the entirety of the video by displaying the numbers in the correct order*

In order to verify the reliability of the Triskell3S architecture and communication protocols, the distributed resolution of the N-puzzle, video was implemented on several sizes of board. The first board is based on a configuration of four lines and four columns (16 bricks) (Figure 4.6). The second experiment has a board of three lines and three columns (nine screens) (Figure 4.7).

The average number of movements of each tile for the resolution of the N-puzzle is equivalent to what can be found in the literature [DRO 93]. The eco-resolution algorithm not having been modified, its values are not detailed here, since they are not relevant to our problem. In each experiment, the relevance of our results is verified by validating the synchronization of the videos in order to form an overall coherent image during the resolution and once the resolution is finished.

Figure 4.7. *Distributed resolution of N-puzzle on the wall of screens of 3*3 IoT-a, each one manipulating videos. The right and lower peripheral screens display the reference video*

The principal contribution of our research is based on the adaptation and implementation of an agent-oriented approach interaction in a real-world environment, in this case on a wall of screens. We will show how these methodologies can be formalized and expanded in the context of connected objects by means of what we have called IoT-a (Internet of Things-agents). The MQTT and D-Bus communication protocols allow us to implement agents in the lower layers of objects, as close as possible to the electronic components. A next step will be experimenting with the reactivity of the agents during the more significant interactions with users of the wall of screens. This step must make it possible to develop new means of interaction with and use of the wall of screens by offering new IoT-a interactions complementary to the *KinectAgent* and *SMSAgent* agents. This step should also lead to experiment with the typology and the optimization of messages to be implemented through predefined protocols MQTT and D-Bus.

4.7. Bibliography

[ABR 09] ABRAS S., Système domotique multi-agents pour la gestion de l'énergie dans l'habitat, Thesis, Institut polytechnique de Grenoble, 2009.

[ATZ 10] ATZORI L., IERA A., MORABITO G., "The Internet of Things: a Survey", *Computer Networks*, vol. 54, no. 15, pp. 2787–2805, 2010.

[BEA 12] BEAUDOUIN-LAFON M., HUOT S., NANCEL M. *et al.*, "Multisurface Interaction in the WILD Room", *Computer*, vol. 45, no. 4, pp. 48–56, April 2012.

[BEL 07] BELLIFEMINE F.L., CAIRE G., GREEWOOD D., *Developing Multi-Agent Systems with JADE*, John Wiley & Sons, Oxford, 2007.

[BER 14] BERGENTI F., CAIRE G., GOTTA D., "Agents on the Move: JADE for Android Devices", *Proceedings of the XV Workshop Dagli Oggetti agli Agenti (WOA 201)*, vol. 1260 of CEUR-Ws, 2014.

[BRD 09] BRDICZKA O., CROWLER J., REIGNIER P., "Learning Situation Models in a Smart Home", *IEEE Transactions on System, Man and Cybernetics*, vol. 39, no. 1, pp. 56–63, 2009.

[CAO 12] CAO J., DAS S.K., *Mobile Agents in Networking and Distributed Computing*, John Wiley & Sons, New York, 2012.

[CAR 06] CARLIER F., BARON M., "Physical object integration and Learning platform in Home Automation", *Proceedings of the International Conference on Mobile Learning (IADIS)*, pp. 147–154, Dublin, 2006.

[CHA 14] CHAPUIS O., BEZERIANOS A., FRANTZESKAKIS S., "Smarties: An Input System for Wall Display Development", *Proceedings of the 32nd International Conference on Human Factors in Computing Systems*, pp. 2763–2772, 2014.

[CHO 12] CHONG N.-Y., MASTROGIOVANNI F., *Handbook of Research on Ambient Intelligence and Smart Environments: Trends and Perspectives*, IGI Global, Hershey, 2012.

[CON 03] FIPA CONSORTIUM, FIPA Communicative Act Library, Specification and FIPA ACL Message Structure Specification, Technical report, 2003.

[COU 08] COUTAZ J., CROWLEY J.L., Plan: intelligence ambiante – défis et opportunité, DGRI A3 report, 2008.

[COU 12] COUTAZ J., CALVARY G., DEMEURE A. *et al.*, "Systèmes interactifs adaptation centrée utilisateur: la plasticité des Interfaces Homme-Machine", in G. CALGARY *et al.* (eds), *Informatique et intelligence ambiante*, Hermes-Lavoisier, Paris, 2012.

[DAS 10] DA SILVA A.R., LAGES W.S., CHAIMOWICZ L., "Boids That See: Using Self-Occlusion for Simulating Large Groups on GPUs", *Computers in Entertainment*, vol. 7, no. 4, pp. 51:1–51:20, January 2010.

[DAV 98] DAVISON B.D., HIRSH H., *Predicting Sequences of User Actions*, AAAI Press, Palo Alto, 1998.

[DBU 16] The DBus home page: https://www.freedesktop.org/wiki/software/dbus, 2016.

[DRO 93] DROGOUL A., DUBREUIL C., "A Distributed Approach to N-Puzzle Solving", *Proceedings of the 12th Distributed Artificial Intelligence Workshop*, 1993.

[DUJ 11] DUJARDIN T., ROUILLARD J., ROUTIER J.-C. *et al.*, "Gestion intelligente d'un context domotique par un SMA", *Journée Francophone des Systèmes Multi-Agents (JFSMA)*, pp. 137–146, 2011.

[GDR 13] GDRI3, GDRI3 Information, Interaction, Intelligence – Thème Systèmes multi-agents, available at: icube-web.unistra.tr/gdri3/index.php/Thème_2_:_ Systèmes_multi-agents, 2013.

[GIL 10] GIL-QUIJANO J.C.H., SABOURET N., "Prédiction de l'activité humaine afin de réduire la consommation électrique de l'habitat", 18ème *Journée Francophone des Systèmes Multi-Agents (JFSMA)*, 2010.

[GLE 14] GLEIZES M., Internet des Objets, Systèmes Ambiants, Report, Soirée Technologie Industrielle, Internet des Objets – Applications et Enjeux, IRIT, 2014.

[GOO 15] GOODYEAR A., HOGBEN C., STEPHEN A., An Innovative Video Wall System, available at: http://www.piwall.co.uk, 2015.

[GRE 06] GREGORI M.E., CÁMARA J.P., BADA G.A., "A Jabber-based Multi-agent System Platform", *Proceedings of the Fifth International Joint Conference on Autonomous Agents and Multiagent Systems*, pp. 1282–1284, New York, 2006.

[GUE 12] GUESSOUM Z., MANDIAU R., MATHIEU P. *et al.*, "Systèmes multi-agents et Simulation", in SEDES F., OGIER J.-M., MARQUIS P. (eds), *Information, Interaction, Intelligence: le point sur le i[3]*, Cépaduès, Toulouse, 2012.

[JAM 14] JAMONT J.-P., MÉDINI L., MRISSA M., "A Web-Based Agent-Oriented Approach to Address Heterogeneity in Cooperative Embedded Systems", *12th International Conference on Practical Applications of Agents and Multi-Agent Systems (PAAMS 2014) Special Sessions*, Salamanca, 2014.

[JAM 10] JAMONT J.-P., OCCELLO M., "Using Hardware/Software Simulation to Design and to Deploy Real World Cooperative Systems", *22th IEEE International Conference on Tools with Artificial Intelligence*, Arras, 2010.

[KIR 15] KIRCHBUCHNER F., GROSSE-PUPPENDAHL T., HASTALL M.R. *et al.*, "Ambient Intelligence from Senior Citizens' Perspectives: Understanding Privacy Concerns, Technology Acceptance, and Expectations", *Proceedings of the 12th European Conference on Ambient Intelligence*, Athens, November 11–13, 2015.

[KUB O8] KUBERA Y., MATHIEU P., PICAULT S., "Interaction-Oriented Agent Simulations: From Theory to Implementation", *Proceedings of the 18th European Conference on Artificial Intelligence (ECAI'08)*, pp. 383–387, Patras, July 2008.

[KUB 11] KUBERA Y., MATHIEU P., PICAULT S., "IODA: an Interaction-oriented Approach for Multi-agent based Simulations", *Journal of Autonomous Agents and Multi-Agent Systems*, vol. 23, no. 3, pp. 303–343, 2011.

[KWA 12] KWAK J.-Y., VARAKANTHAM P., MAHESWARAN R. *et al.*, "SAVES: A Sustainable Multiagent Application to Conserve Building Energy Considering Occupants", *Proceedings of the 11th International Conference on Autonomous Agents and Multiagent Systems*, Valencia, June 4–8, 2012.

[LAM 12] LAMPKIN V., LEONG W.T., OLIVERA L. *et al.*, *Building Smarter Planet Solutions with MQTT and IBM WebSphere MQ Telemetry*, Vervante, Springville, 2012.

[MAM 12] MAMIDI S., CHANG Y.-H., MAHESWARAN R., "Improving Building Energy Efficiency with a Network of Sensing, Learning and Prediction Agents", *Proceedings of the 11th International Conference on Autonomous Agents and Multiagent Systems*, Valencia, June 4–8, 2012.

[MAT 13] MATHIEU P., PICAULT S., "The Galaxian Project: A 3D Interaction-Based Animation Engine", *Proceedings of Advances on Practical Applications of Agents and Multi-Agents Systems (PAAMS'2013)*, Salamanca, May 22–24, 2013.

[MOZ 05] MOZER M.C., "Lessons from an Adaptive Home", in COOK D., DAS R. (eds), *Smart Environments: Technologies, Protocols, and Applications*, pp. 273–294, John Wiley & Sons, New York, 2005.

[NAN 15] NANCEL M., PIETRIGA E., CHAPUIS O. *et al.*, "Mid-Air Pointing on Ultra-Walls", *ACM Transactions on Computer-Human Interactions*, vol. 22, no. 5, pp. 21:1–21:62, August 2015.

[NOR 88] NORMAN D.A., *The Psychology of Everyday Things*, Basic Books, New York, 1988.

[PAT 04] PATON C., Development of a Message Oriented Interaction Layer for Agent Communication, Computer Science Technical Reports, University of Bath, 2004.

[PAW 14] PAWLOWSKI K., KURACH K., SVENSSON K. *et al.*, "Coalition Structure Generation with the Graphics Processing Unit", *Proceedings of the 2014 International Conference on Autonomous Agents and Multi-Agent Systems (AAMAS'14)*, Paris, May 5–9, 2014.

[PIE 11] PIETRIGA E., HUOT S., NANCEL M. *et al.*, "Rapid Development of User Interfaces on Cluster-driven Wall Displays with jBricks", *Proceedings of the 3rd ACM SIGCHI Symposium on Engineering Interactive Computing System (EICS '11)*, New York, 2011.

[RAM 11] RAMCHURN S., VYTELINGUM P., ROGERS A. *et al.*, "Agent-based Homeostatic Control for Green Energy in the Smart Grid", *ACM Transactions on Intelligent Systems and Technology*, vol. 2, no. 4, p. 35:1–35:28, 2011.

[RAO 04] RAO S.P., COOK D.J., "Predicting Inhabitant Action Using Action and Task models with Application to Smart Homes", *International Journal on Artificial Intelligence Tools*, vol. 13, pp. 81–100, 2004.

[RIH 13] RIHAWI O., SECQ Y., MATHIEU P., "Impact des politiques de synchronization dans les simulations réparties d'agents situés", 21ème *Journée Francophone des Systèmes Multi-Agents (JFSMA)*, Lille, 2013.

[SAB 14] SABOURET N., JONES H., OCHS M. *et al.*, "Expressing Social Attitudes in Virtual Agents for Social Training Games", *Proceedings of the Second International Workshop on Intelligent Digital Games for Empowerment and Inclusion at IUI 2014 (IDGEI 2014)*, Haifa, 2014.

[SER 93] SERVAT D., DROGOUL A., "Combining Amorphous Computing and Reactive Agent-based Systems", *Proceedings of the 12th Distributed Artificial Intelligence Workshop*, Seattle, 1993.

[STE 05] STERLING B., *Shaping Things*, MIT Press, Cambridge, 2005.

[SUB 14] SUBAGDJA B., TAN A.-H., "On Coordinating Pervasive Persuasive Agents", *Proceedings of the 2014 International Conference on Autonomous Agents and Multi-agent Systems (AAMAS '14)*, pp. 1467–1468, Paris, May 5–9, 2014.

[UPT 12] UPTON E., HALFACREE G., *Raspberry Pi, User Guide*, John Wiley & Sons, New York, 2012.

The Visualization of Information of the Internet of Things

5.1. Introduction

Today, many things are created with intelligence in mind, for example a building equipped with control sensors for electricity, water consumption managed from the web or the automation of taxi services via the cellular network and GPS, where the use of applications has encouraged the development of an interconnected world. Mobile devices connect larger and larger numbers of people, objects and services.

This observation highlights the concept of smart objects and smart devices, which are digital electronic devices that are connected to each other thanks to different types of networks and protocols (Bluetooth, 3G, 4G, Wi-Fi) that lead to the creation of a collection of Smart Objects (smartphones, Smart Cars, Smart Homes, Smart Cities and Smart World) which are visible each time we connect to the Internet.

Five initiatives of Smart Devices are attractive according to Stankovic [STA 14]: Internet of Things (IoT), Mobile Computing (MC), Pervasive Computing (PC), Wireless Sensor Networks (WSNs) and Cyber-Physical Systems (CPS). Research on the IoT, PC, MC, WSN and CPS is concentrated primarily on technologies like real-time computing, machine learning, confidentiality, security, signal processing and Big Data among other things. At the same time, Smart Vision involves diverse domains of

Chapter written by Adilson Luiz Pinto, Audilio Gonzales-Aguilar, Moisés Lima Dutra, Alexandre Ribas Semeler, Marta Denisczwicz and Carole Closel.

science as well as the creation, management and use of interconnected smart objects linked to the Internet.

According to Kopetz [KOP 11]: "... connecting physical objects to the Internet allows remote access to data from sensors used to monitor the physical world remotely. A Mashup of data captured extracted from diverse web sources leads to new services which go beyond the services provided by an isolated system. The IoT is based on this vision".

A final part of the concept is the overview of connected objects, which simultaneously offers a telescope and microscope for the parts of the world that are invisible to people, machines and physical objects [GRE 15]. The IoT, by searching for meaning in movements between objects, people, animals, vehicles, air currents, viruses, etc., creates the practical and conceptual framework of a connected world.

In this regard, Chung proposes integrative service systems and providers of solutions from the Internet of Things to institutions [CHU 15]. Network operators providing communication services, infrastructure for transporting data, production of software, equipment for the manufacture of GPS chips, Wi-Fi, sensors, portable devices, as well as integrated material devices for data placement are all included in the concept of the Internet of Things.

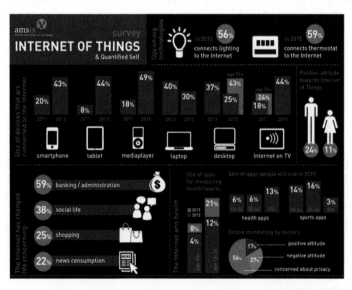

Figure 5.1. *The Internet has changed daily life.*
Source: https://www.ams-ix.net/newsitems/87

"The Internet of Things, by linking a huge quantity of devices that communicate with each other, us and the Internet, takes shape. There is still a certain skepticism, but the fact is that the Internet has a central place in people's lives and this development will not stop. The Internet simplifies and improves many daily activities," says Job Witteman, CEO of AMS-IX (https:// ams-ix.net/newsitems/87).

AMS-IX interviewed 1,100 consumers to create a complete image of the future daily use of the Internet of Things by consumers around the world.

When people were asked about the number of devices they own, what became apparent was that the use of smartphones, tablets and televisions with an Internet connection is increasing. Two years ago, 20% of people surveyed had used a smartphone. Today, 43% of people surveyed own a smartphone and 44% have a tablet. Moreover, 18% of people questioned earlier had a television with an Internet connection, while today the rate is 38%.

These responses show an expected growth of connectivity in televisions, thermostats, lighting and multimedia readers in the next two years. Over the course of two years, the percentages of connected devices will change in the following way: 44% of televisions will be connected; connected thermostats will increase from 24% to 59%; lighting from 24% to 56% and multimedia readers will increase from 32% to 49%.

The devices that are used least are office computers (from 37% to 25%) and portable computers (from 40% to 30%). By observing the different age groups, it quickly becomes apparent that the over 55 group is ahead of the curve with its use of televisions which have an Internet connection.

The Internet continues to play an essential role in the daily life of populations. It has radically changed day-to-day activities. The people polled said that the Internet has changed how they perform their administrative and banking operations (59%), allows them to maintain social contacts (38%), make purchases (25%) and follow the news (22%). The Internet has a more significant impact on social contacts for women (23%) than for men (16%), while men listen to music and follow information in a different way from women.

Our work intends to show that visualization is fundamental in the interfaces of connected objects and that it plays an essential role with visual analytics in the Internet of Things.

5.2. Internet of Things

The Internet of Things (IoT) is an extension of the current Internet in which many objects, sensors and devices, referred to as "objects" are connected and integrated with each other. This integrated group of objects can be considered a whole with the capacity to act and work collectively.

The IoT is aimed at linking people and objects everywhere and at any time [PER 15]. This connection allows people to interact with their objects and for these objects to interact with each other. According to the McKinsey Global Institute [MCK 15], these interactions permit the creation of systems that monitor the state and actions of connected objects and machines. Moreover, they can also monitor the physical world, people and animals. The IoT is a scenario in which objects, animals or people are equipped with unique identifiers, with the possibility of automatically transferring the data over a network without the need for human intervention [ATZ 10].

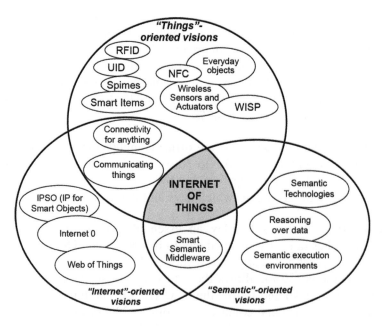

Figure 5.2. *Oriented vision view of objects, the Internet and semantics. Source: [ATZ 10]*

For Singh, Tripathi and Jara [SIN 14], the IoT represents the convergence of the Internet with RFID (*Radio-Frequency IDentification*) technology, sensors and smart objects. RFID is a technology that allows communication between devices by using chips for the wireless transmission of data. With this type of automation, any device can be identified using an RFID tag.

The IoT can be considered a new Internet revolution. According to Vermesan and Friess [VER 15], the IoT allows for the integration of the physical world into the virtual world by using the Internet as a means of communication and exchange of information. The main goal of the IoT is therefore to bring the physical world closer to the digital world. The numerous devices connected to the Internet will produce an enormous quantity of data collected in the physical world [ANA 13]. According to Wang *et al.* [WAN 13], the collection of this raw data must be processed effectively, namely, the data must be analyzed in a way that allows for the extraction of information that is valuable or representative for people. The convergence of these networks of devices will produce a large amount of raw data, which will have to undergo processing by humans or machines, so that useful and practical information can be extracted. This development has the potential to change the way people see their "objects" and the Internet itself.

The IoT can be applied in diverse domains, with the potential to generate numerous business opportunities as well as opportunities to improve the way in which certain services within these domains are currently available to the public. One can imagine that IoT could have a significant impact on people's daily lives. As we can see in Figure 5.3, the IoT can be applied to many different domains. These domains are separated into i) industrial (industrial domain); ii) smart cities (Smart City Domain) and iii) health and well-being (well-being and healthcare), which are then subdivided in turn, as proposed by Borgia [BOR 14].

Su *et al.* [SU 14] estimate that the application of the IoT in diverse domains could help people to improve their quality of life by providing advantages in many areas, from health to agriculture. Moreover, it should be stressed that there is room for improvement and the creation of innovative initiatives.

For example, one of the domains where a lot of research has been done recently is that of smart cities. Smart cities enable the creation of intelligent environments thanks to the use of technologies to ensure different functions within the town such as intelligent transportation. The field of smart cities is one of the domains that presents challenges that need to be met to generate improvements. For the McKinsey Global Institute [MCK 15], towns can be used as one of the principle centers of innovation, since they carry within them several questions that can be developed.

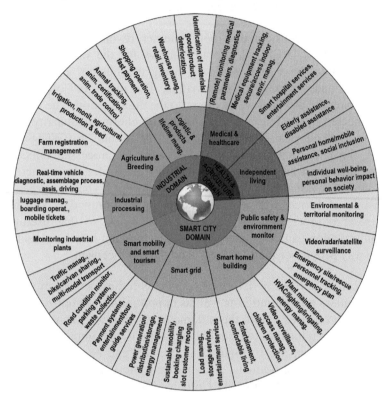

Figure 5.3. *Fields of application for the IoT andconnected applications. Source: [BOR 14]*

It should be specified, however, that the fields of application in Figure 5.3 are not an exhaustive list. Moreover, not all of them have the same level of maturity [BOR 14]. Barnachi estimates that, for the IoT to become entirely functional, it is necessary to resolve the problems linked to the diversity, volatility and ubiquity of data that make the data processing a difficult task [BAR 12]. There are also challenges linked to software infrastructure, and consequently new processing and data visualization services must be developed to support applications in an evolving and interoperable environment. Certain authors think that one of the most significant problems of the IoT is the interoperability of information [IER 13].

In the opinion of Gubbi *et al.*, to produce an accurate visualization of the data coming from the IoT, the interfaces of devices must be user-friendly and intuitive for everyone [GUB 13]. Moreover, any user interaction with the environment will require adequate visualization software, which will highlight detection mechanisms such as those for interpreting the data collected [SIN 14].

5.3. InfoVis and DataVis in the Internet of Things

The current forms of collaboration between science and computing form e-Science which, with Big Data innovations and technologies, focuses on the intensive use of data produced by computer simulation. The value that can be extracted from this data requires changes in the form of analysis of visual data (DataVis) and requires that technological innovations allow an interaction for visualizing the large volume of data generated in real time (data streaming), such as climatic predictions, astrophysical predictions and trade flows (InfoVis). In this context, visual analysis of data is closely linked to the Internet of Things, to Big Data and data visualization.

Big Data technologies are aimed at providing tools for visualizing data collections originating from a multitude of sources and areas of knowledge such as those of physics, astronomy, business, environmental monitoring, disaster and risk management, security and analytical engineering. The collection of data from these domains is incomplete, heterogeneous and assumes different data formats (videos, texts and metadata) which requires mass storage and also requires rapid analysis of what has been collected. Data originates from sensors and electronic devices, including content from the Cloud. In this way, Big Data technologies can provide revolutionary discoveries for science and visual analysis industry in the future.

Big Data is a technological and human phenomenon which is attempting to resolve current problems in the intensive use of data according to its volume, speed, veracity and variety [CHE 15].

In this context, the science of data appears as a field of study on the work, techniques and software of data search using algorithms that focus on the extraction and visualization of large quantities of data. *Data Science* provides the skills and knowledge necessary to meet the challenges of Big Data. It is concerned with the use of data and facilitates the visualization process of changes in the domains of health, business and insurance as well as the efficient management of energy resources. It is based on traditional techniques such as *Data Mining, Machine Learning, Visual Analytics*, high-performance Cloud computing, *Parallel Computing* and the collection of information [MAS 15].

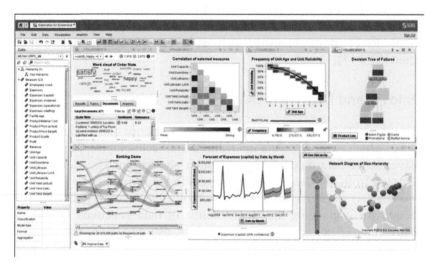

Figure 5.4. *Analytical system for data. Source: [SAS 16]*

The use of technologies for data visualization has a fundamental importance for data scientists who are concerned with identifying models, trends and relationships in collections of data by using Big Data technologies. Directly related to visualization is *visual analytics*, an area of study that involves the use and interactive visual analysis of large amounts of complex data (dataset) that represents the analytical process and requires a high degree of surveillance and man-machine interaction.

In parallel, Data Science is a science involving data that defines the current forms of visual analytics with precision. In short, Data Mining and Cloud Computing can be considered the first stages of the transformation of Big Data.

Data Science is a new field of exploration, that will be able to resolve the current and future problems related to Big Data. This combination offers numerous possibilities for data scientists, engineers and information technology companies. It also offers mathematical possibilities for discovering new algorithms for data visualization [CHE 15].

Data Science requires a form of systems thinking, by combining a creative approach to generally pragmatic problems. This approach is exemplified by a way of thinking like that of a civil engineer, combined with that of a visual designer. Planning is essential for working with Data Science since it requires many resources and different approaches to obtain results [VOU 14].

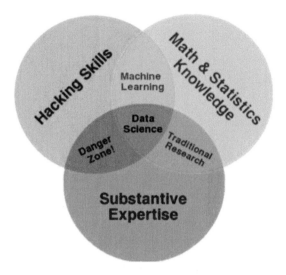

Figure 5.5. *Diagram of scientific data. The Venn diagram of data is under Creative License as Derivative-Non-commercial license. Source: [CON 10]*

5.3.1. *Visual analytics in the context of the Internet of Things*

Visual analytics is the science of reasoning that relies on the use of interactive visual interfaces for its representation. Currently, the data is produced at an incredible speed and the capacity to collect and store data has increased at a faster pace than the capacity to analyze the latter [WON 04]. Over the last few years, a large number of automated data analysis methods have been developed. Nevertheless, the complex nature of the problems require the inclusion of human intelligence at an early stage of the data analysis process.

The visual analytics method allows decision-makers to combine flexibility, creativity and basic human knowledge with the storage and processing capacities of computers in order to resolve complex problems. The use of advanced visual interfaces allows for direct interaction between humans and computers, by reducing data analysis efforts, and by allowing them to make effective decisions in complex situations.

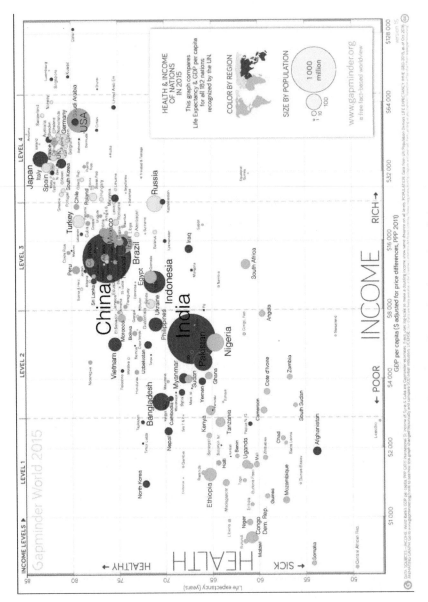

Figure 5.6. *Example of visual analytics.*
Source: [ROS 12]. Creative Commons License

Visual analytics combines the automatic analysis of data with interactive visualizations. This definition could include the processes of creation and the use of software and technical analysis for the visualization of data. Visual analytics is described as a science that uses analytical reasoning to facilitate the process of comprehension of visual interfaces. It's an interactive process that involves the collection of information, data processing, the representation of knowledge, interaction and decision-making. It involves the description of large quantities of data, whether this be scientific, medico-legal or originating in businesses from heterogeneous sources of data. It takes care to combine high-level computing, such as Big Data, with human perception and cognition. To do this, methods related to automated analysis are used, such as *Knowledge Discovery in Databases* (KDD), statistics and mathematics, while the human side entails perceptive capacities for relation and decision-making, which makes visual analytics a future area for research, deeply linked to the tasks of the Data Scientist [THO 05].

The fields of application for visual analytics pertain to a range of segments related to diverse research practices: physics; astronomy; business; environmental monitoring; risk management; security; biology; medicine; analytical engineering, etc.:

– physics and astronomy include applications such as the visualization of flows, fluid dynamics and molecules, nuclear sciences, astrophysics, acquisition and collection of data on the universe. Volumes of unstructured data at the Big Data scale originate from different directions of space orbits and cover the frequency spectrum, thus forming continuous flows of terabytes of data than can be recorded and analyzed by supercomputers. In this case, the quantity of data is so high that it surpasses human capacity for comprehension. Thanks to data analysis techniques such as KDD, astronomers can discover new phenomena, useful relationships and knowledge about the universe, etc. A visual analytics approach can help separate the pertinent data from the noise and help identify phenomena in the massive and dynamic flow of data;

– business: the financial market with its different actions, obligations, raw materials, stock indexes, currencies and money, generates a large amount of data every second, and thus accumulates large volumes of data over the years;

– environmental monitoring: reports on events and climatic and meteorological conditions. This is a domain that involves the collection of large quantities of data from satellites and sensors. These sensors capture data on the changes in the climate that have taken place during the day at successive intervals. This data accumulates in terabytes. The applications in this domain, are, on the one hand, instantaneous visualizations (snapshots), that is to say, instantaneous images of a situation or an event in real time, and on the other, they generate sequences of past events and predictions for the future. This makes it possible to analyze certain phenomena and identify certain essential factors in their development and helps the decision maker take the measures necessary such as the global reduction of carbon dioxide emissions to reduce global warming. Applications used for modeling and climate visualization can cover any time intervals possible, including daily meteorological predictions made at short intervals, which make up the basis of complex visualizations of climate change. This action can also extend to predictions over periods covering thousands of years;

– risk management makes it possible to predict environmental disasters, helped by previous visualizations of climate change, and to take the necessary measures, such as the construction of physical barriers or the evacuation of populations. These scenarios can include natural disasters or meteorological conditions (floods, rogue waves, volcanic eruptions, storms, fires or endemic diseases) but also technological disasters caused by humans such as accidents, traffic accidents or pollution. Thus, visual analytics can help predict the extent of the damages and makes it possible to define appropriate and effective strategies for the affected area;

– security: the field of application in this sector is vast and covers the protection of information systems against cyber-terrorism, as well as network security. In these domains, challenges reside in obtaining all of the information in order to find correlations;

– biology and medicine: the areas of research of biology and medicine can offer a wide variety of applications. For example, computer tomography and 3D ultrasonography in the medical field. Another emerging application is bioinformatics which offers many possible applications for visual analysis, such as the Human Genome Project, with three billion base pairs of the human genome. Other domains are emerging, such as proteomics (the study of proteins in a cell), metabolomics (the systematic study of chemical fingerprints of specific cellular processes), or combinatorial chemistry which has already identified tens of millions of compounds and which is further expanding each day. Traditional visualization techniques cannot deal with this volume of data, the new visual analysis methods turn out to be, on the other hand, more effective for analyzing data in these contexts;

– analytical engineering covers the entire range of processes related to civil engineering, for example the physical processes of construction or the automobile industry, for example the resistance of vehicles. Another application in the automobile industry is the simulation of a car accident, where the image of a car is represented as a grid of hundreds of thousands of points and the accident is simulated by a computer.

Alongside the experiments mentioned, visual analysis of data uses images to represent information. It requires an interdisciplinary knowledge of mathematics, computer programming, visual perception and cognitive sciences. It therefore makes it possible to explore the theories and practices of visualization, where it is involved in the acquisition of knowledge in different fields of application.

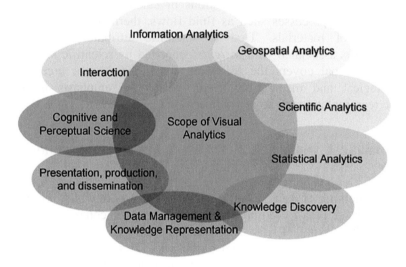

Figure 5.7. *Fields related to visual analytics.*
Source: based on [KEI 08]

Visual analytics can be considered as a sort of direct application of the techniques that consolidate scientific data. It combines visualization solutions for the disciplines that arise from the intertwining of studies on information visualization (InfoVis) and scientific visualization (sciVis). Its main objective is to provide techniques and tools that oversee the analysis and extraction of knowledge from visual interfaces. It is aimed at developing the capacity to transform complex data in interactive and significant visualizations. Its practitioners learn the fundamental principles of data design and the typology of data visualization [TEL 15]:

– InfoVis: this is the visualization of information. It is by nature interdisciplinary and covers computer graphics, geography and the information sciences. InfoVis has great potential to improve access, processing and management of large quantities of information. So the role of this technique is to reduce the data in a unique environment, which simplifies the analysis. Among the tools used for this type of representation are scientific charts [BOR 02].

– SciVis: this type of visualization was initially used to refer to visualization as being part of the scientific computing process (the use of modeling and computer simulation in scientific research and engineering). SciVis attempts to respond to the problems related to the increasing quantity of data generated by digital simulations of calculations originating from diverse physical processes such as fluid flows, thermal convection or the deformation of materials. Tied to scientific attributes, this type of visualization is associated with *insights* generated by scientific simulations. Its recurring goal covers the following three phenomena: architectonic, meteorological and medical-biological, where the emphasis is put on realistic representations of large volumes of data and surfaces [POS 03].

5.4. Analytical visualization in the context of the Internet of Things

Visual analytics of data is characterized by the way in which it represents and summarizes data by combining several types of codes in descriptive graphics. This type of display uses quantitative synthesis computer technologies, such as graphics and dashboards. Thus, Big Data represents unique challenges for data visualization, since it formulates and provides information extracted from data [FEW 08].

It is also possible to deduce that visual analytics can be a method that combines the techniques of automated analysis with interactive visualizations to make comprehension of data effective, and facilitate reasoning and decision-making when the analysis of large amounts of complex data is carried out [KEI 10].

Visualization is at the heart of the system, not only as a means of communicating the value of data or the results of an analysis, but it is, additionally, used more and more to monitor processes in other disciplines, such as data management and Data Mining (DataMining).

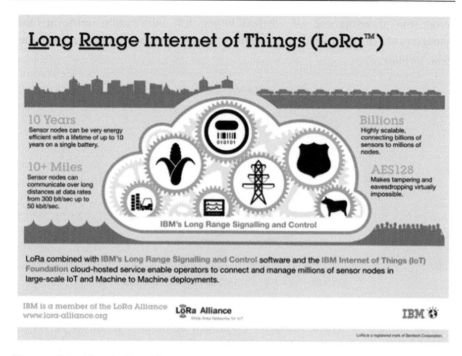

Figure 5.8. *The LoRa Alliance® guarantees the interoperability and flexibility technique IoT applications. Source: https://www.flickr.com/photos/ibm_research_zurich/ 16371812028/in/photostream/. Image under Creative License by-nd 2.0*

Its usefulness is expanding due to the excess of information that circulates on the web throughout the world. In 2014, more than 210 million e-mails, 4 billion SMS and 90 million tweets were sent per day. In Europe, Media Monitor is a system that automatically determines what is covered in the media and collects around 2,500 documents that come from new sources: media portals, government sites and press agents. It processes around 80,000 to 100,000 articles per day, in 43 languages [KEI 10].

Visual analytics can be seen as an approach that combines visualization, human factors and data analysis. Figure 5.9 shows the research domains related to visual analytics.

Visual analytics includes, in addition to visualization, data analysis and human factors, cognition and perception. It also plays a key role in man-machine communication, and facilitates the decision-making process. Visual analytics involves connected domains: the visualization of information, computer graphics (computer graphics-interfaces) and data analysis, which favors the development of methodologies for information retrieval, the management of data and the representation of knowledge, as well as Data Mining.

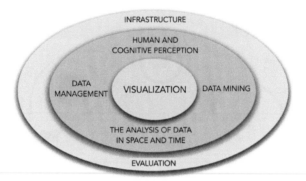

Figure 5.9. *Domains related to visual analytics.*
Source: [KEI 10]. Translation by the authors

The process of visual analytics combines automated methods of visual analysis with human interaction, with the goal of acquiring knowledge from the data.

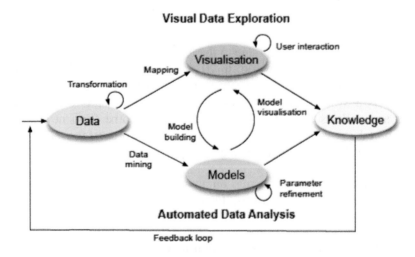

Figure 5.10. *The visual analytics process.*
Source: [KEI 10]. Translation by the authors

Figure 5.10 provides a general overview of the steps (presented in the shape of ovals) and transitions (arrows) of the visual analytics process.

The process involves two fronts:

– the pre-processing and transformation of data to obtain different representations leading to numerous application scenarios. Heterogeneous sources of data must be integrated into visual analysis methods and applied in an automated way. Consequently, this first step is used to obtain different forms of representation and exploration (as indicated by the arrow in the processing of Figure 5.10). Other typical tasks include pre-processing, cleaning data, standardization, consolidation and integration of heterogeneous data sources;

– the application of automated methods for visual analytics, following data processing, the analyst can choose between the application of visual analysis methods or automated analysis. When an automated analysis is used for the first time, data extraction methods are applied to generate models of primary (original) data. Once the model is created, the analysis must evaluate and refine the models that can be produced by interaction with data.

Visualizations allow analysts to interact with automated methods and modify the parameters and the selection of analysis algorithms. The model of the display can then be used to evaluate the results of the models generated. Alternating between visualization and automated methods is the visual analytics process, which leads to ongoing improvement and verification of preliminary results.

Incorrect results identified during the intermediate state can be considered an initial state, leading to better results and greater confidence in the analysis. When the visual scanning of data is done as a first step, the user can confirm the hypotheses generated during the automated analysis of the intermediate phase. The user's interaction with the visualization is necessary to reveal useful information, for example, zoom in on different areas of data or different perspectives in the analysis of visual data.

Thus, the display of the results can be used as a guide to construct automated models of analysis. In conclusion, knowledge of the processes of visual analytics and automated analysis, as well as the interactions with visualizations, precedes models by human analysts [THO 05].

5.5. Conclusion: the relevance of the use of visualization in the Internet of Things

The saying "a picture is worth a thousand words" underlines the popular conception of Big Data and takes on its full meaning in the contemporary world where users and businesses generate enormous quantities of stored data. This data has no value; it cannot be displayed in a simple way that is accessible to any user for rapid and effective decision-making. Although in the last few decades, graphics have been used more and more to visualize data from businesses, visualization technologies have been improved to match current needs for mobility, where Big Data reveals a new user profile from the world of business [LIE 13]. The integration of technology and the optimization of data visualization allows for the display of key information through graphics, tables, charts, etc. It then becomes possible to draw conclusions in a simple and visual way, which is essential for businesses so they can make decisions in real time, improve their performance, learn about domains and anticipate problems in order to prevent them from posing a real risk to the company [MIG 14].

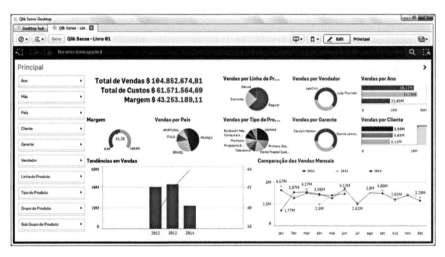

Figure 5.11. *Generation of visual analytics with Qlik® Sense Desktop. Source: [FRE 14]*

This type of data management for visual analytics also includes tips for optimization:

– infographics for a better presentation of the data in a simplified form, which gives users the possibility of streamlining the content, showing what is essential for decision-making. The possibilities are infinite, from the display of heart rate on a watch to the graphic ease of use of a mobile telephone monitoring system;

– digital marketing to be developed via the concept of connectivity through smartphones, since in a future world visualization will be applied to microprocessors and combined with household equipment. With this in mind, certain businesses have begun to make their implementation known, as was the case with Panasonic showing the programming and temporality of its electric and electronic systems with intelligent usage systems, by voice-based systems or remotely by mobile;

– the analysis of social networks that can be used to control towns or vehicles, and control of systems in high-risk areas or under circumstances that are dangerous for humans. It can also be used to control the traffic of information, analyze its density and proximity, as well as identifying the central points of connectivity;

– GPS systems that also facilitate the availability of location data via the system of geographical maps. They can serve as a basis for remote sensing systems for floods, earthquakes, climatic processes, etc.

In conclusion, our work shows that in the future visualization will play an increasingly essential role in the Internet of Things since it is at the heart of interfaces, analysis and decision-making.

5.6. Bibliography

[ANA 13] ANANTHARAM P., BARNAGHI P., SHETH A., *Data Processing and Semantics for Advanced Internet of Things (IoT) Applications: Modeling, Annotation, Integration, and Perception*, ACM Press, New York, 2013.

[ATZ 13] ATZORI L., IERA A., MORABITO G., "The Internet of Things: A survey", *Computer Networks*, vol. 54, no. 15, pp. 2787–2805, 2010.

[BAR 12] BARNAGHI P. *et al.*, "Semantics for the Internet of Things: Early Progress and Back to the Future", *International Journal on Semantic Web and Information Systems*, vol. 8, no. 1, pp. 1–21, 2012.

[BOR 02] BÖRNER K., CHEN C. (eds), *Visual Interfaces to Digital Libraries*, Springer Verlag, Berlin, 2002.

[BOR 14] BORGIA E., "The Internet of Things vision: Key Features, Applications and Open Issues", *Computer Communications*, vol. 54, no. 1, pp. 1–31, 2014.

[CHE 15] CHEN L.M., SU Z., JIANG B., *Mathematical Problems in Data Science: Theoretical and Practical Methods*, Springer International Publishing, Amsterdam, 2015.

[CHU 15] CHUNG J.M., Internet of Things and Augmented Reality Emerging Technologies, Course curriculum, Yonsei School of Electronical and Electronic Engineering, Seoul, 2014.

[CON 10] CONWAY D., "Venn Diagram", available at: http://drewconway.com/zia/2013/3/26/the-data-science-venn-diagram, 2010.

[FEW 08] FEW S., "With Dashboards: Formatting and Layout Definitely Matter", available at: https://www.perceptualedge.com/articles/Whitepapers/ Formatting_and_LayouMatter.pdf, 2008.

[FRE 14] FREITAS A.R., "O que é o Qlik Sense e o Qlik Sense Desktop?", available at: http://www.guiatecnico.com.br/gt/?p=480, 2014.

[GRE 15] GREENGARD S., *The Internet of Things (Essential Knowledge)*, MIT Press, Cambridge, 2015.

[GUB 13] GUBBI J. *et al.*, "Internet of Things (IoT): A vision, architectural elements, and future directions", *Future Generation Computer Systems*, vol. 29, no. 7, pp. 1645–1660, 2013.

[IER 13] IERC, IoT Semantic Interoperability: Research Challenges, Best Practices, Solutions and Next Steps, IERC AC4 Manifesto – Present and Future, 2013.

[KEI 08] KEIM D. *et al.*, "Visual Analytics: Scope and Challenges", in SIMOFF S.J., BÖHLEN M.H., MAZEIKA A. (eds), *Visual Data Mining: Theory, Techniques and Tools for Visual Analytics*, Springer-Verlag, Berlin, 2008.

[KEI 10] KEIM D. *et al.*, *Mastering the Information Age Solving Problems with Visual Analytics*, Eurographics Association, Goslar, 2010.

[KOP 11] KOPETZ H., "Internet of Things", in KOPETZ H. (ed.), *Real-time Systems*, Springer, New York, 2011.

[LIE 13] LIEBOWITZ J., *BigData and Business Analytics*, CRC Press, Boca Raton, 2013.

[MAS 15] "Master In Data Science", available at: https://www.city.ac.uk/courses/postgraduate/data-science-msc, consulted January 15, 2016.

[MCK 15] MCKINSEY GLOBAL INSTITUTE, "The Internet of Things: Mapping the Value Beyond the Hype.", available at: http://www.mckinsey.com/insights/business_technology/the_Internet_of_things_the_value_of_digitizing_the_physical_world, 2015.

[MIG 14] MIGUÉLEZ J.A., "Seis recomendaciones para optimizar la visualización de datos em la empresa", *Business Intelligence*, available at: http: //www.dataprix.com/empresa/recurso-it/business-intelligence/6-recomendaciones-optimizar-visualizacion datos-empresa, 2014.

[PER 15] PEREIRA C. *et al.*, "Big Data Privacy in the Internet of Things Era", *IEEE Computer Society*, vol. 17, no. 3, pp. 32–39, 2015.

[POS 03] POST F.H., NIELSON G.M., BONNEAU G.-P., *Data Visualization: The State of the Art*, Springer, New York, 2003.

[ROS 12] ROSLING H., "Gapminder World", available at: http://www.gapminder.org/downloads/world-pdf/, 2012.

[SAS 16] SAS VISUAL ANALYTICS, "Data Visualization Software that Offers Full-size Power for Any Size Budget", available at: http://www.sas.com/content/sascom/en_za/software/business-intelligence/visual-analytics/_jcr_content/par/styledcontainer_7674/par/contentcarousel_5b29/cntntcarousel/textimage_9afd/image.img.png/14577 23467804.png, 2016.

[SIN 14] SINGH D., TRIPATHI G., JARA J.A., "A Survey of Internet-of-Things: Future Vision, Architecture, Challenges and Services", *IEEE World Forum on Internet of Things (WF-IoT)*, Seoul, March 6–8 2014.

[STA 14] STANKOVIC J.A., "Life Fellow, Research Directions for the Internet of Things", *IEEE Internet of Things Journal*, vol. 1, no. 1, pp. 3–9, 2014.

[SU 14] SU X. *et al.*, "Adding Semantics to Internet of Things", *Concurrency Computation: Practice and Experience*, vol. 27, no. 8, pp. 1844–1860, 2014.

[TEL 15] TELEA A., *Data Visualization Principles and Practice*, CRC Press, Boca Raton, 2015.

[THO 05] THOMAS J., COOK K., *Illuminating the Path: Research and Development Agenda for Visual Analytics*, IEEE-Press, Los Alamitos, 2005.

[VER 15] VERMESAN O., FRIESS P. (eds), Internet of Things IoT Semantic Interoperability: Research Challenges, Best Practices, Recommendations and Next Steps, European Research Cluster on the Internet of Things, IERC, March 24 2015.

[VOU 14] VOULGARIS Z., *Data scientist: The Definitive Guide to Becoming a Data Scientist*, Technics Publications, Basking Ridge, 2014.

[WAN 13] WANG W. *et al.*, "Knowledge Representation in the Internet of Things: Semantic Modelling and its Applications Automatika", *Journal for Control, Measurement, Electronics, Computing and Communications*, vol. 54, no. 4, pp. 388–400, 2013.

[WON 04] WONG P.C., THOMAS J., "Visual analytics", *IEEE Computer Graphics and Applications*, vol. 24, no. 5, pp. 20–21, 2004.

The Quantified Self and Mobile Health Applications: From Information and Communication Sciences to Social Innovation by Design

6.1. Introduction

Connected objects and portable screens are being integrated into our everyday lives little by little. They are becoming smaller, increasingly ergonomic and less and less perceptible when worn on the human body. They can collect physiological, behavioral and geo-localized data. As a result, a culture of a body that is more equipped with technological objects that make it possible to collect, store and visualize personal information about the self is developing. To that extent, we have entered into "the culture of the Quantified Self" [LAM 14], based on the self-measurement of personal parameters and interconnection between portable screens, connected objects and social networks. Numerous objects with increasingly elaborate devices accompany athletes or simple citizens who want to gather data on themselves. Chris Dancy is a striking example of this practice. This North American resident collects large quantities of data about himself day and night. Between 2010 and 2013 he lost a large amount of weight as a result of the impact of *biofeedback*, which allowed him to use information technologies linked to each other via the Internet of Things. His physical transformation was displayed on the multiple platforms that make up his digital identity in such a way that it constitutes a paradigmatic example of

---

Chapter written by Marie-Julie CATOIR-BRISSON.

the possibility of modifying the body using the Quantified Self. His daily use of connected objects synthesizes both the promises and the concerns of connected health, home automation, and enhanced reality used for the purposes of prevention or even behavioral prediction. This unique experiment is interesting to analyze in order to grasp what the integration of information technologies into our daily lives entails. The analysis of Chris Dancy's use of objects is aimed at answering the following questions: how do connected objects transform the relationship between the individual, his body and its representation, how do they move the human-machine relationship toward more online social interactions and lead to a form of spectacularizating communication of his own data? To understand the multiple issues that this problem raises, an interdisciplinary approach combining the analytic tools of semiotics, design and the anthropology of communication is proposed.

The qualitative analysis involves a corpus made up of data collected from the multiple platforms that make up Chris Dancy's digital identity. Although our observation phase stretches over three months (February–May 2015), the data collected falls within a longer period of time (2010–2015) in order to analyze Chris Dancy's physical transformation. The corpus is made up of different types of supports: photos and texts published on socio-digital networks (Facebook, Instagram, Twitter), visual data from two applications used by Chris Dancy (FitBit and Existence), the discourse on his relation to connected objects published on his two web sites (www.servicesphere.com and www.chrisdancy.com) as well as in interviews (in particular those done for the magazine *Mashable*) and slideshows of his conferences about his information technology experiments on his Slideshare channel. Within the framework of this study, it was necessary to make choices about the large corpus of data collected. The selection was made using criteria salient to the problem by choosing in particular the data that demonstrates the factitive dimension of the objects used by Chris Dancy and the specific relationship that he was with them.

With regard to the methodology, the analysis of the corpus is done in three levels. The first concerns the study of the staging of his physical transformation on socio-digital networks and the social interactions that generate this staging. The second deals with the aesthetic dimension of the representation of data on "health" applications from a semiotic perspective, to characterize the information design of applications, the factitive dimension of connected objects used by Chris Dancy, as well as the value system at the heart of interfaces. This analysis is completed by an intermedial approach that puts information design and data visualization into historical perspective. Finally, there is a look at Chris Dancy's vision of interaction design aimed at understanding how this paradigmatic example crystallizes a particular relationship to information technologies and is part of a trajectory of technologies that should be interrogated critically.

The study is broken down into four parts. First, it involves revisiting the definition of certain terms related to the Quantified Self and m-health in order to characterize the information technologies used by Chris Dancy. The second part presents the results of the selective analysis, emphasizing Chris Dancy's transformation. The third part concerns the factitive dimension of the devices used by Chris Dancy, in particular the value system at the heart of his relationship with his portable technologies and his vision of interaction design. From an anthropological perspective, it also involves placing this phenomenon within the conception of information technologies and embedded computing. The fourth part provides a critical perspective built around this particular case. It opens up to a more global reflection on the bioethical, institutional and socio-economic challenges raised by connected objects used in the healthcare sector. It also looks at how the development of connected objects and ubimedicine transforms the triadic relationship between patients, doctors and public and private health institutions. Finally, the article proposes other paths to consider [GRA 13] for m-health technologies based on the anthropology of communication and social innovation by design.

6.2. The evolution of interfaces and connected objects toward anthropotechnics

6.2.1. *From e-health to the "Quantified Self"*

First of all, it seems necessary to define and distinguish between the terms often associated with the promotional discourse accompanying the development of self-measurement devices and mobile applications in the field of health and well-being. E-health, m-health and the Quantified Self are often used together, which contributes to the confusion for users, even though they cover very different data processing processes and practices, as depicted in Figure 6.1.

According to the European Commission, the term e-health (or e-Health) "refers to all of the technologies and services for medical care based on ICT" [COM 09]. It includes a variety of diverse practices, from telemedicine to information systems for healthcare professionals. The use of the term has been trivialized to the point where it just refers to "everything that contributes to the digital transformation of the healthcare system" [CON 15]. M-health (or m-Health) is an extension of e-health focused on mobility using portable information technologies and devices connected to a mobile network. It simultaneously includes medical practices and public health supported by mobile devices as well as monitoring and surveillance of patients via communicating measuring devices. Finally, the measurement of the self (or Quantified Self) "refers to a group of varied practices which all have the common characteristic of measuring and comparing variables relating to someone's way of life" [CNI 14]. The development of the Quantified Self is related to the development of the Internet of Things. The Quantified Self movement has been growing at such a high rate since 2011 that it gives the impression of being innovative and unprecedented. However, self-measurement has been a common practice since the introduction of home scales and thermometers in the 19th Century. The novelty of the Quantified Self is not the act of measuring the self but rather that the data collected is being shared via the Internet of Things. Consequently there is a major difference between m-health and the Quantified Self in terms of the collection and access to data. Indeed, in m-health, it is the healthcare professionals who ask patients to collect the data, which remains between them and their patients. With the Quantified Self, it is the individual who takes the initiative to measure his or her personal data and communicate it to others, in particular via socio-digital networks.

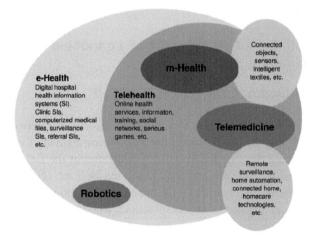

Figure 6.1. *Distinction between Telehealth, e-health, m-health and Telemedicine (Connected health. Livre blanc du Conseil national de l'ordre des médecins, 2015, p. 9)*

At his point, we can ask ourselves if the daily use of multiple portable sensors also falls within the domain of health. For the OMS, health encompasses a "state of physical, mental and social well-being." We can therefore decide that Chris Dancy's use of information technologies, which cannot be reduced to the monitoring of his physiological data, falls outside the healthcare field. It seems more pertinent to consider his use of portable information technologies as relevant to "anthropotechnics" which are defined as: "the art or technique of extra-medical transformation of the human being through intervention on his own body" [GOF 06].

6.2.2. *Anthropotechnics and the information ecosystem of Chris Dancy*

The illustrations below make it possible to visualize the different devices worn: connected bracelet, armband and glasses, camera and portable camera, physiological sensors and connected portable screens that record his movements and geo-localize him.

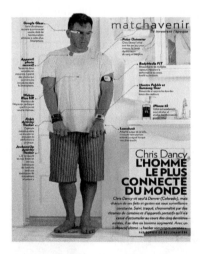

Figure 6.2. *The devices[1] worn every day by Chris Dancy. Photograph published in Paris Match, July 2014. http://www.parismatch.com/ Vivre/High-Tech/L-homme-le-plus-connecte-du-monde-577862*

---

1 To understand how these sensors work, we can refer to the "Guide to Self-Tracking Tools". Source: http://quantifiedself.com/guide/.

In addition to these portable devices, other technologies are integrated into Chris Dancy's domestic space[2] which refer back to the field of home automation. For example, to measure his sleep, he combines a sensor next to his bed with the bracelets that he wears on his wrist. He also uses heat and movement sensors in the room. The lights and music are programmable remotely for creating a particular mood, especially when he returns home after a business trip. In this way, Chris Dancy deploys an entire complex information and communication ecosystem made up of both the connected objects that he wears and the sensors integrated into his household furnishings. He performs the daily recording of his physiological and biometric data in order to analyze and visualize it on the three computer screens that he uses in his office. All of the devices are interconnected via Wi-Fi and the data circulates from one interface to another (in particular from smartphone screens to computer screens).

His office is a very distinctive space. Several types of object coexist: three flat screens and multiple sensors (a cube sensor for example) are mixed with books (in particular ones by Warhol) and decorative objects. His wall is covered with wood panels on which he has created a collage of press clippings, photos, phrases for meditation and multiple objects. There are also statues next to the digital devices. He uses two different seats: an ergonomic chair for working on these three screens, and a wooden chair decorated with feathers and multiple sculptures, which recalls those of the great Native American chiefs.

Figure 6.3. *Chris Dancy in his office: cultural and technological hybridization. Photograph published on the website of the magazine Mashable in August 2014, http://mashable.com/2014/08/21/ most-connected-man/#0TM6VmdLGkq1*

2 Chris Dancy presents his objects and connected domestic space in a video made by Bianca Consunji and Evan Engel, for the online magazine *Mashable*, 2014. Source: https://www.youtube.com/watch?v=qdCQUHxVxfk.

Thus, this piece illustrates the fetishistic relationship that Chris Dancy maintains with his information technologies: it is a sort of datacenter from which he controls his data. This space makes up the cornerstone of his self-monitoring system and is characterized by a strong cultural and technological hybridization.

Another room attached to the office is reserved for recharging all of the devices that he uses every day. It is equipped with a USB hub with more than 30 ports to recharge the hundreds of connected objects that he uses.

6.2.3. *Connected objects as the heirs of ubiquitous computing*

Although the development of connected objects seems new, it's necessary to place it within the history of information technologies. Chris Dancy's connected objects are communicating objects, characterized by "their capacity to mutually recognize one another" (LCN, 2002). The development of communicating objects follows "a general trend of the spreading and burying of technologies" [DEM 02]. It is based on two major changes: "the digital convergence of information technologies and the common "mobilification" of these objects [PRI 02]. By melting into the domestic environment, communicating objects tend to disappear into the environment to create ambient communication, building a symbiotic relationship between Chris Dancy and his surroundings. This particular interaction has already been considered in works on "ubiquitous" computing and "the attentive environment" [WEI 91]. It leads to a transformation of the relationship between the user and the interfaces, since the multiple interconnected objects make up a diffuse interface, "submerged transparently in the environment" [PRI 02].

From a middle perspective, we can observe a competition and synergy between computer screens and smartphones around which Chris Dancy's communicating objects gravitate. Connected objects give the smartphone a central role because designers of connected objects and mobile applications prefer its screen, which is adapted for mobility. There is also a complementarity between the telephone and computer screens, if you consider synchronization between these two screens, and the circulation of mobile media [CAT 12] from connected objects to the smartphone then to the computer. On the other hand, the portable screens that accompany Chris Dancy daily (such as his smartphones or his connected bracelets) behave like personal digital assistants that recall the PDAs that appeared at the end of the 1980s.

From a semiotic and socio-cultural point of view, Chris Dancy's screens are simultaneously "action" and "contact" screens [LAN 10]. His portable screens also belong to the category of "intimate screens" [TRE 14], whose principle characteristics are mobility and experience. Chris Dancy himself divides his data into "little data" and "experiential data" highlighting the passage of Big Data to the individual scale. The paradox of these intimate screens lies in the fact that they are interconnected with the public sphere, via the network of the web: in this way, they mediate Chris Dancy's personal data and share it well beyond his intimate sphere. All of this leads to questions about the role of screens and connected objects in his relationship to others and to his own body.

6.3. Factitive dimension and value system at the heart of Chris Dancy's relationship with his information technology

We can consider Chris Dancy's connected objects as "factitive objects" [DEN 05] and analyze the way in which they shape his behaviors and social interactions. By offering him recommendations based on the data collected, they have participated in the modification of his body and his social life. The factitive dimension of his connected objects rests in their capacity to provide *biofeedback* in real time, which leads him to modify his behavior. This quantification of his daily life is then transformed into *individu-data* [MER 13].

6.3.1. *The progressive development of the figure of the enhanced human in socio-digital networks*

Three photographs posted on Facebook play a part in the dramatization of Chris Dancy's physical metamorphosis from 2010 to 2013. This display constitutes a spectacularization of his body, as part of a storytelling approach [SAL 08]. Beyond the transformation of Chris Dancy's face, resulting from his weight loss, it is interesting to observe the evolution of the attributes of his persona. Starting in 2011, he wore glasses with very visible black contours. Then, in 2013, he acquired a Google Glass that he wears for all his media appearances, to the point that it has become central to his "enhanced" human persona.

The value of Chris Dancy's metamorphosis as a result of his anthropotechnics appears in the comments of his "friends" on the 2013 photo: they describe him as "handsome" or even "looking great." The change therefore involves his relationship to others and himself, with a self-improvement approach.

2009   2011   2013

Figure 6.4. *Dramatization of Chris Dancy's metamorphosis on Facebook (profile photo)*

6.3.2. *Information design and data-visualization: the case of Fitbit and Existence*

We can then analyze the mobile applications used by Chris Dancy by exploring both information design and data-visualization. This second level of analysis involves a double dimension: aesthetic and semiotic. In the application Fitbit, the data collected is visualized in the form of daily graphics: bar graphs for the step counter, line graphs for heart rate, donut charts to evaluate sleep quality. The application also measures the calories burned by the user. In the application Existence, the user can explore his or her *timeline* to analyze his or her daily activity via donut charts, and optimize his time, at the same time that he gets feedback on the same daily activity. The application appears to be designed with an "ethic of compassion" in the vein of contemplative computing[3] [SOO 11].

These two interfaces are nomadic, since the data can be consulted on a smartphone and a computer via the Internet. They seem to represent two information design trends in the Quantified Self: the first is created from a design centered on performance, while the second claims to be contemplative design, meant to distinguish it from disruptive computing. In the two applications, we can see the recurring use of donut charts, which are the dominant representation in many applications based on data-visualization. If we place the information design of these applications into a historical perspective, we observe that graphics are just a re-actualization of the principles of graphic semiology [BER 67] adapted to contemporary visual culture. What is new, however, is that these are personal messages

---

3 Chris Dancy often uses the term *contemplative technology*, citing the work *The Distraction Addiction* by Alex Soojung-Kim Pang (2013).

addressed to the user by the computer system to congratulate or encourage him. This significant detail is an important feature for users of these objects that have been transformed into a kind of life coach.

The data that Chris Dancy collects on himself is partly processed and visualized by health/well-being applications, as well as also through manual processing and data visualization with different software programs (in particular Evernote, Spreadsheets and Google Calendar). These systems are familiar to him because he was for more than twenty-five years in the information technology sector, working for businesses. The illustration below makes it possible to visualize the complexity of the information systems that he deploys.

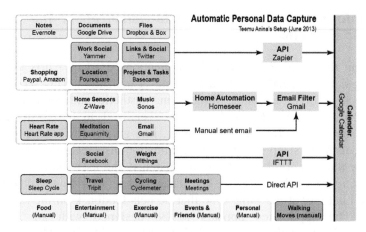

Figure 6.5. *"Diagram of the workflow" created by Teemu Arina for Chris Dancy's blog. http://www.servicesphere.com/blog/2013/12/5/ explaining-my-quantified-self-or-coming-out-of-my-data-close.html*

We can observe that a part of the data collection is carried out automatically by the different software programs to which they are connected. Another part of the data, concerning food, entertainment, physical exercise and social life, is entered by Chris Dancy himself every day.

6.3.3. *Animism and anthropomorphism: a particular relationship to connected objects*

We can finally look at Chris Dancy's discourse on the intimate relationship he maintains with his self-measurement information technologies. On his website and blog, he presents himself as "the most connected human in the world" and a "mindful cyborg". His stated goal is to map his existence, with the help of the hundreds of sensors, devices, applications and services that he uses every day. The goal of his approach appears clearly in the slideshows that he shares on his Slideshare[4] channel, especially in "Existence: the human information system."

Figure 6.6. *The concept of "fluid self." Screen capture of a slideshow by Chris Dancy on Slideshare. http://fr.slideshare.net/chrisdancy/the-human-information-system-byod-wearable-computing-and-imperceptible-electronics*

Chris Dancy distinguishes three types of data that make up the "self" called the fluid self: soft data, hard data and core data. Thus, for him, our behavior is made up of the multiple facets of our digital experience: an isotopy emerges from the terminology used to describe the data at the scale of the "self" and that of the computing language including the image of a networked body, pictured as a complex and transparent information system. On socio-digital networks, Chris Dancy claims that his connected objects help him to be "a better human being" and allow him to better understand himself. Moreover, he states on Twitter, "It's not about your data, it's about your identity". His discourse is aimed at convincing others that the

---

4 Chris Dancy's presentation channel. Source: http://fr.slideshare.net/chrisdancy.

Quantified Self used for the purposes of benevolent self-monitoring is accomplishing the Socratic quest for self-knowledge.

In the interviews and images that he publishes on Facebook, he creates the image of a hybrid body that references those of science fiction, by intermingling spirituality, embodied technology and invisibilization of digital interfaces in the environment. Moreover, he declares that having multiple pieces of information on the environment surrounding him, thanks to his connected classes is "like being the Terminator!". Likewise, a Facebook profile image (Figure 6.7) is a photomontage in which Chris Dancy's face appears with a fluorescent necklace around his neck, a computer component embedded in his cheek. A "friend" commented on this image by comparing Chris Dancy to Captain Kirk from the film *Star Trek*.

Figure 6.7. *"The mindful cyborg": the hybrid body and science-fiction imagery. Screen capture of a post by Chris Dancy on Facebook*

Another profile image is a portrait created by Aaron Jasinski in 2012, in which Chris Dancy is shown in profile, facing a robot (Figure 6.8). He holds a robot mask in his hand while the robot holds a mask with his face. The portrait is titled *"The Real You"*, which stresses the fact that Chris Dancy represents himself as being half-human and half-robot. It is also interesting to note that a comment posted by a close friend who seems to know him well refers to "this cyborg wearing your face." The comment seems to signify that

Chris Dancy feels closer to a robot than a human, and has constructed a hybrid identity of an "enhanced" human.

Figure 6.8. *"The real you": portrait created by Aaron Jasinski. Screen capture of a post by Chris Dancy on Facebook*

All of these elements are disseminated on the multiple platforms which make up his fragmented digital identity, contributing to the promotion of an image of the interaction between humans, information systems and the environment. Chris Dancy moreover proposes the term *"Innernet"* for picturing a possible future in which the individual interacts with the environment through feedback which the connected objects *on* and *around* him send back to him. In this vision of interaction design, the body and the environment become interfaces, and identity is defined by information.

Chris Dancy also specifies the orientation of his approach to self-measurement by distinguishing between Big Brother and Big Mother. According to him, the former system accumulates data about individuals in order to control them, while in the latter system, the collection of data on oneself is done for the purposes of taking control of the data by and for himself. The use of the expression "Big Mother" is a strategy aimed at reinforcing the benevolent dimension of this self-monitoring, as it brings to mind imagery linked to maternity.

A comment published by Chris Dancy on Instagram regarding an image from the application Existence emphasizes the emotional relationship that he maintains with his information technologies. There is a dog peaceably lying down with the following phrase "Are you feeling better today? You weren't yourself recently." Chris Dancy comments: "It knows me". Attributing a human capacity for knowledge, comprehension and compassion to the application underlines the emotional link that he has with this application, characterized by emotional design [NOR 12] focused on both the behavioral and reflexive levels. This user-friendly interface can only reinforce the burying of technologies and the machine's different layers of calculation, at the same time that it creates one with the user. In addition, the touch screens with which Chris Dancy interacts every day are constructed on "a progressive analogy between human sensoriality and mechanical sensibility, bringing machines closer and closer to bodies" [MPO 13].

Thus, the analysis of these several texts and images published online by Chris Dancy make it possible to understand his relationship with his connected objects, his vision of interaction design and the value system at the heart of this relationship with his devices: physical performance, self-esteem, hybridization (cultural, technological, physical), incorporation and invisiblization of portable information technologies are its main axes. Chris Dancy's experience constitutes a paradigmatic example of the ideology of transhumanism, whose relationship to technology is made up of a mixture of animism[5], anthropomorphism and refers to science fiction imagery.

Chris Dancy's vision of interaction design oriented toward *contemplative computing, calm technology* and the attentive environment falls within the continuum of Mark Weiser's ubiquitous computing project. Chris Dancy's experience seems to crystallize an emerging trend in our contemporary era, in which the relationship to a world mediated by interfaces is generalized, while the interfaces *invisibilize* into the environment.

Chris Dancy's work on the dialectic of the values of connected objects allows an updating of "an axiology and an alignment with social values" [BEY 12]. We can thus picture the way in which connected objects embody the contemporary zeitgeist by offering a relationship to the world mediated by interfaces. The use of Big Data at the individual scale seems to realize the original dream of information technologies and to communicate and accomplish the biopolitical project of cybernetics on the scale of the human body.

From a socio-cultural point of view, Chris Dancy's portable information technologies transform him into an "interfaced man" centered on a logic of "auto-reading-writing between brain and screen, and auto-regulation of the

---

5 Dominique Boullier's (2002) developing analysis of anthropological issues is at the heart of our relationship with communicating objects, and in particular the animist relationship that we hold with them.

body" [REN 14a, REN 14b]. The Quantified Self is part of the fantasy of the "datafication of life itself" [CUK 14] and constitutes an ideology that is widely publicized to the point of becoming dominant in the doxa. However, this orientation of personal information technologies is only one possible path [GRA 13] for Big Data that needs to be interrogated.

In the case of Chris Dancy, it is also necessary to mention the ambiguity in his discourse between the promotion of the tools he uses (by incorporating the slogans of the applications in his slideshows presented in his conferences and on Slideshare) and his user experience, knowing that he is an expert in information technologies. Confusion arises about the genre of the discourse (reflective, advertising) and in the roles that he incarnates (consultant, witness, user) which come close to conflicts of interest. This confusion is part of the experiential marketing movement, in which consumers turn into spokespeople for brands (applications in this case).

6.4. Critical perspective and avenues for reflection for reconsidering the use of connected objects and mobile applications in the field of health

At the end of this analysis, it seems apt to go beyond the particular case of Chris Dancy in order to develop reflections on the ethical, institutional and socio-economic challenges related to the use of connected objects and Big Data in the healthcare field. If Chris Dancy's case is still exceptional today, connected objects are used more and more widely by citizens. To respond to this growing demand, a market in information technologies specializing in health is developing exponentially to the point of becoming a challenge for society which involves not only both the community of researchers in information and communication sciences, engineers and designers, and doctors, but also all citizens more generally. This final part therefore has the goal of critically considering connected objects and applications available on the market, but also envisaging other possible technologies and the way in which they could be created within an ethical and sustainable dynamic, based on a logic of human-centered design.

6.4.1. Ethical and social issues related to data governance

"It always comes with a price…" this saying from the series "Once upon a time" seems appropriate for thinking about the passage from spectacularization of data to the problem of data governance. From an ethical point of view, a tacit contract is created – without it being spelled out – between the citizen and manufacturers of information technologies.

As Dominique Cardon emphasizes in his most recent work, "the subject's confrontation with the quantification of his behaviors is promoted as an instrument for constructing identity, a personal *benchmark*" [CAR 05]. In the case of Chris Dancy, his use of Big Data is linked to the three steps of his self-improvement project: lose weight, stay in shape, and most recently be "Zen". The use of information technologies then falls within the utopia of transparency that the digital permits, one which maintains the idea that "cross referencing of data done by the user himself" allows him to be "administrator of his own data" by becoming a "collector-interpretor" [CAR 05, p. 78]. The user thus becomes master of his data in the face of "state surveillance" and "the instrumentalization of the market." However, this vision is illusory since, "when individuals take control of their data, they do it in a context of asymmetry of information and the absence of alternatives" [CAR 05, p. 79]. Thus, personal information technologies and Big Data at the individual scale offer many opportunities, but they also pose questions beyond the governance of data. This problem exists in front of a judicial void since there is currently no law that regulates circulation and cross-referencing of data. Furthermore, the digitization and cross-referencing of health data leads to the anonymization of patients, whose data is accessible on the network [MIC 15]. It is at this level that the problem of data regulation by public and legal institutions that protect citizens is raised. "Quantification practices in the health field favor individual micro-management of health to the detriment of a more collective understanding. They make individuals into entrepreneurs of themselves who are responsible for their good or bad health habits, and can distract attention from the environmental and socioeconomic causes of public health problems" [ROU 14].

This system of self-monitoring raises the problem of the digital traceability of personal data. Connected objects call into question the interpretation and use of data from a bioethical, political and socio-economic point of view. Recording physical performances according to the norms and standards defined by manufacturers in connected "health" raises the question of the role of mediation by public institutions and healthcare actors (especially doctors) to frame the use of data and prevent the commercialization of health. This problem entails "algorithmic" [ROU 13] thinking. While the development of Big Data draws on a concept of transparency supposed to empower the individual, data collection techniques in fact threaten the empowerment of citizens. Big Data is initiating a new regime of visibility presented as a neutral goal, while it removes the collectively negotiable neutral perspective for the benefit of mechanical representation, producing norms with which it is impossible to negotiate.

This is why the exponential development of connected objects, much like that of nano- and biotechnologies, involves "the redefinition of the relationships between civil and technological society" [JAR 14, p. 327]. A reflection on the future of human and individual identity in the world of Big Data seems necessary, at a step in the development of biotechnologies where it is still possible to question them. Used only for commercial purposes, these technologies could end with a society of control, via technologies integrated into the privacy of the body. Within the context of privatization of health insurance, connected objects could lead to a commodification of biological and medical formulas, which would be harmful to the less performing of us.

The development of Big Data is also becoming a challenge in terms of marketing, since it permits traceability of consumption on the web and the possibility of predicting the behavior of consumers. This situation is a blessing for the large industrial technology groups: it can lead to a new form of voluntary subjugation, in which the individual becomes the tool of individualized marketing.

The ethical and social questions raised by the massive use of connected objects in the field of health are many. It is first necessary to consider the development of "ubimedicine" – a term suggested by Dr. Nicolas Postel-Vinay (Hôpital Européen Georges Pompidou, Paris) to refer to "what could be a medical practice based on the reception and analysis of health data collected voluntarily by the user in multiple times and places"[6]. This neologism reinforces the fact that this practice was developed outside of the traditional institutional frameworks (such as the consulting room or the hospital room) and follows in the footsteps of ubiquitous computing. We cannot deny that the development of "connected health" is the result of a triple evolution: sociological, technological and politico-economic, made manifest by the lightning-fast expansion of the market of connected objects oriented toward health or well-being. However, this development cannot only remain in the hands of the sector's manufacturers. It needs to be supervised by doctors, guarantors of the protection of medical

---

6 Conseil national de l'ordre des médecins (2015), Santé connectée: de la e-santé à la santé connecté, Le Livre blanc du CNOM, January 2015, p. 12.

confidentiality, and public institutions for the regulation of patient data, as well as the patients themselves; all these must be considered in this digital transformation of the health sector that involves the area of public policy. The White Paper published by the *Conseil national de l'ordre des médecins* (CNOM), raises two major questions. Doctors insist on the necessity of asking: "to what extent can they [applications] be considered medical devices? Do we have to stipulate specific rules for the protection of the data collected?"[7]. We can complete this line of questioning by also asking who will be guarantors of data protection. Who can access it and for what purpose? This questioning refers to the problem of regulating the circulation of patient data that involves all healthcare actors (doctors, patients, public and private institutions) who participate in the patient's course of care. This therefore involves rethinking the French social solidarity model, by considering the way in which the doctor/patient relationship is transformed by the use of connected objects and mobile health applications.

6.4.2. *The doctor-patient relationship transformed by connected objects and mobile health applications*

With connected objects and mobile health applications, doctor-patient communication is mediated by digital interfaces and the patient data is communicated beyond the medical sphere, especially in the case of the digitized medical file and sharing over a network. Patients become consumers of information services, which profoundly transforms their relationship with healthcare professionals, but also the way in which data related to health is produced.

As we saw in the beginning of this study, it is necessary to distinguish medical devices from anthropotechnics. The pattern of consultation for these two types of technology is radically different. In the first case, a patient has consulted a doctor who performs a diagnostic and proposes treatment including the collection of certain data using digital and non-digital tools. In the second, it's the consumer who initiates both the diagnosis and treatment since he self-evaluates using anthropotechnics and information services to transform his body by himself. The goals are not the same because medicine relies on a code of ethics that falls within a legislative framework. The Big Data used in health thus calls into question two pillars of medicine: medical

---

7 Conseil national de l'ordre des médecins (2015), Santé connectée: de la e-health à la connected health, Le Livre blanc du CNOM, January 2015, p. 11.

confidentiality and "the search for the best benefit/risk balance for health in keeping with the patient's autonomy" [GOF 13, p. 101]. This is how the doctor-patient relationship is evolving toward a "professional-client relationship" or even a "service delivery relationship" [GOF 13, p. 96].

In addition, the positivism that is exacerbated by technologies that are more and more user-friendly also maintains the illusion of the possibility of making a medical diagnosis which does not require any human mediation, or even that is more performing than a healthcare professional. This therefore questions the role of scientific mediation, and also brings up the problem of legal responsibility for the interpretation of health data by a third party that does not belong to medical professionals. When it comes to information and communication with the patient, promotional discourse maintains confusion about the medical goal of "health" applications, with a generalized trend with manufacturers in the information technology sector toward medicalization of connected objects, or at least the claim of a health benefit. There is thus a problem with the deficit of information about m-health, even though consumers, lost in the jungle of connected objects available on the market, are waiting for advice and recommendations from healthcare professionals. This problem of misinformation resembles the arrival in the market in the agro-food industry of nutraceuticals, which were marketed as food having properties similar to medications that provided providing health benefits.

The use of digital devices in the field of health "could be an effective resource for cooperation between the person and his or her doctor, more generally with the healthcare professionals that oversee care" [CON 15, p. 33]. Nevertheless, it is more necessary than ever for doctors to participate in the conception and setting up of digital medical devices, as well as in the reflection on the regulation of health data, in dialogue with public and private institutions, and in the social interest of the patient. This also entails offering a digital education for doctors and patients, which returns society more generally to the problem of digital literacy and digital humanities. On the one hand, doctors need to be trained, within the framework of their curriculum, in the use of digital medical devices (comprising connected objects and mobile health applications). "The training must deal not with the tool but with its integration of ethics and professional standards into the medical practice itself, for the benefit of the patient." On the other hand, this digital education also involves patients, who must be encouraged to "promote use respectful of rights and freedoms, confidentiality and the protection of personal data" [CON 15, p. 34].

Thus, connected objects and mobile health applications could constitute a socially useful complement to consultation in many cases, involving the "monitoring of a metabolic disruption such as diabetes, a diet designed for weight loss, assistance with therapeutic education, support or maintenance of autonomy or monitoring of physical and athletic activity" [CON 15, p. 34]. It is nevertheless necessary to define "between the doctor and the patient, a framework of "appropriate use" of the application or connected object during its integration into the field of care and treatment" [CON 15, p. 34]. This is why doctors and more generally healthcare professionals need to address the problem of connected health in order to offer solutions adapted to their needs and to those of their patients. This social project imagines "the realization of a double evaluation, combining value-in-use and medico-economic value" [CON 15, p. 33]. Although the mediation of doctor/patient communication is increasing with the development of the use of communicating objects and digital interfaces, it is nevertheless necessary for digital mediation to be framed by human mediation, by placing the healthcare professionals at the heart of the debate over the use of these new devices.

6.4.3. *The necessity of considering the point of view of doctors and healthcare professionals*

Several reports and investigations published at the national[8] and international[9] scale attest to a growing preoccupation with the development of connected and mobile health on the part of healthcare professionals. These dialogues have given way to analysis of the use of digital devices within the community of doctors and patients, some of which have led to concrete propositions that seem to us important to consider. To develop reflections on the use of connected objects and health applications serving the needs of patients and doctors it is indispensable to listen to these propositions of healthcare professionals. Among the available professional

---

8 On the national scale, we can cite two reports targeted in particular at the use of digital devices by doctors: "Usages Numériques en santé: 2ème baromètre sur médecins utilisateurs de smartphone en France", Observatoire Vidal, May 2013 and "Baromètre annuel sur les usages digitaux des professionnels de santé", CESSIM-Ipsos, 2014, a study involving 2,800 doctors and pharmacists.

9 On the international scale, OMS has already engaged with the field of m-health, by publishing "mHealth New horizons for health through mobile technologies", Global Observatory for eHealth series – Volume 3, OMS, 2011.

literature, the propositions of the CNOM are particularly interesting, because they synthetically present workable solutions to accompany the surge of connected objects within the framework of the doctor/patient relationship, in particular within the context of office visits.

At the moment, there is no certification in France concerning the use of applications and connected objects that are not recognized as medical devices. In addition, the principal goal of CNOM's propositions is to better inform patients about the functions and conditions of use of these devices. To provide the means of carrying out this project of education in the use of digital devices, CNOM has developed six propositions [CON 15, pp. 6–7]. The first proposition is aimed at "defining proper use of mobile health in the service of doctor-patient relationships" which entails defining an ethical framework for integrating m-health devices into medical care. From this perspective, another proposition reinforces the necessity of "watching over the ethical use of connected health technologies" by attracting attention to economic models based on the valorization of patient data and risk threatening national solidarity. When it comes to the designers and manufacturers of connected health, CNOM advises "promoting appropriate, progressive and European regulation" and to "pursue scientific evaluation" by experts independent of the sector's manufacturers. It seems important to point out the importance of the necessity for applications and connected objects to meet a certain number of standards in order to be recognized as medical devices, which entails both the challenges of regulation but also interoperability between devices. For patients, the major challenge of m-health is to "develop digital literacy" especially concerning the mastery of the advanced functions of digital devices in terms of confidentiality and the protection of personal data. Finally, it also involves initiating a "national e-health strategy", and now m-health, which involves French and European political decision-makers in the healthcare field to clarify the governance of health data and respect the confidentiality of citizens' personal data and the necessity of their consent for their use. This last proposition asks us to consider digital devices "not as an end but as a group of methods making it possible to improve access to care, the quality of treatment, the autonomy of patients". It clearly emphasizes the necessity of initiating a debate between actors in public policy, doctors and patients, who must participate in the deliberations over the use of their health data.

Taking users into account (doctors and patients especially) then encourages a move toward a qualitative approach and a form of research making it possible to offer solutions adapted to the needs of each of the stakeholders of an m-health project.

6.4.4. *Envisaging other paths for m-health technologies based on the anthropology of communication and social innovation by design*

Interaction design's anthropological approach can contribute to the development of a critical reflection on the dominant models of interaction design only focused on technological or economic innovation. This approach constitutes a constructive "techno-critique" [JAR 14] for offering alternative models of digital technologies in the field of health. Digital technologies and their identification as "new" technologies must be analyzed as sociotechnical and political devices. From this perspective, the works of anthropologist Lucy Suchman constitutes an enlightening avenue for research for highlighting the necessity of an approach that takes the quality of the interaction between human, digital interfaces and the environment into account. In an article dedicated to the links between anthropology and digital design, Lucy Suchman insists on the need to develop a "critical anthropology of design that contributes to the emergence of a critical practical technique" [SUC 11, p. 16]. From this perspective, it is pertinent to analyze connected objects and health applications not only as "intelligent machines" to which humans delegate a part of their social practice, but also as an "embodied form of social practice" [SUC 11, p. 8]. This approach makes it possible to take into account the interaction that takes place between users and their material goods and social environments, by considering the environment as a situational mediation that plays a role as important as technological mediation, in the experience offered to the user.

The analysis of human-machine interaction by the anthropology of communication and design can thus contribute to the reflection on "human ecology" understood as the sum of interactions that take place between humans and their environments, both natural and artificial at the same time [FIN 15] and to the development of concepts, methods and tools in the field of social and digital innovation. To develop the analysis of the interaction between digital devices, users and their environment, it is necessary to take into account the fact that connected objects, like computers, are "physically incarnated and incorporated in a context in such a way that their capacities and their limits depend on a physical substratum, an environmental situation" [SUC 11, p. 7]. Works coming from the anthropology of communication are interesting for the design of digital health technologies, and in particular the possibility of returning to the works of Erving Goffman in *Rites d'interaction*, which have been considerably transformed by digital technology in the case of connected health. The concept of frame proposed by this researcher is pertinent for rethinking the doctor/patient relationship, the relationship of the patient with his or her own health data, or the communication between patients and health institutions. This deep qualitative analysis of the rite of interaction that is the consultation would make it possible to offer new forms of design in the space of a medical office, more adapted to communication mediated by digital interfaces, and the imperatives of therapeutic education of patients in the use of digital medical devices.

It would therefore seem wise to develop, based on a critical approach to the dominant digital devices in the m-health sector, a form of interdisciplinary research capable of contributing to the reflection and the conception of innovative digital technologies from a social and environmental point of view which is rooted in the problem of informational ecology. Researchers in humanities and social sciences need to be involved in this reflection, by the side of specialists in medical disciplines, in order to participate in the emergence of other models of digital technologies and interaction design in the field of m-health.

The criticisms and fears related to the current path of technology can be seen as the "symptom of a crisis of confidence requiring the setting up of structures of dialogue between businesses, publics and profane powers" [JAR 14]. Design understood as a discipline of the project, is a pertinent approach that is complementary to critical theories of information and communication sciences. Human-centered approaches (human-centered design) are particularly interesting for organizing a dialogue on a major topical issue that concerns every citizen. Human-centered design is not limited to design centered on the user, even if taking use into account plays an important role. Human-centered design is "research on what can support and reinforce the dignity of human beings and the way in which they live in diverse social, economic, political and cultural circumstances" [BUC 01, p. 35]. In that respect, "the quality of the design is distinguished not just by technical skill of execution or the aesthetic vision, but above all by the moral and intellectual goal toward which technical and artistic skill is directed" [BUC 01, p. 26]. In this respect, co-design with stakeholders (doctors, patients, public and private health institutions) would make it possible to offer solutions adapted to specific cultural features (both local and national) and to the needs of the users by taking into account social, economic and ethical imperatives. This approach to technologies through design and the anthropology of communication thus makes possible the movement from a logic of technological innovation toward a logic of social innovation.

Social innovation is a concept that is reappearing today in the scientific debate, notably in the field of design. This dimension is not really new, insofar as the central problem of design is to explore ways in which to improve the world's livability. It reconnects rather with the essence of design, understood as a discipline of the project, in particular in as the works of Bauhaus, Victor Papanek, or Alain Findeli show. Several definitions of social innovation exist and, as much as the field has been under debate since the beginning of the 2000s, we can nevertheless note three characteristics common to social innovations [VIA 15]. They are not primarily commercial and are located on the side of the common good because their beneficiaries are collective. They are created with the goal of responding to social needs. And they rely on new forms of governance in which the beneficiaries are involved in a participative way, which transforms social relationships. In this respect, social innovation includes a sociopolitical dimension through the recognition of the power of individuals and communities to plan and act. It entails rethinking traditional project management methods by involving new actors, beyond the industrial and economic sectors, particularly as a result of co-design methods.

The return of the social in design seems above all to signal the desire of certain researchers and designers to break from industrial design and be part of a logic of social innovation, anchored in the problems of contemporary society. Indeed, reflection on social innovation by design is connected to that of sustainable development and falls within the context of the current transition that we are living in our hypermodern societies. As Ezio Manzini [MAN 07] emphasizes, the mission of design is to support the way in which individuals redefine their existence in individual self-directed or collective projects. The role of the designer is therefore to create conditions favorable to collaborative work in order to support the process of social and societal change.

Social innovation by design is thus clearly distinguished from technological innovation by putting the concerns of the individual and his or her aspirations at the center and by relying on a dynamic of horizontal, transversal and participative research. This form of research is located on the border of observational and interventional research, by simultaneously offering a close observation of the uses of an existing device and a new proposition for conception of a device adapted to the needs of the stakeholders in a project. Alain Findeli has notably proposed a model of design research called "project research" [FIN 15], in which the research is conducted within the framework of a design project that constitutes the field of research. Project research in design is particularly interesting for contributing to the reflection on alternate forms of digital technologies in the service of the needs of doctors and patients in the field of m-health. At the crossroads of humanities and social sciences, design moreover makes it possible to imagine transformations that are not only technological, but also social, cultural and communicational brought about by the digital.

Project research in social and digital innovation and by design can thus contribute to the criticism of the dominant models of digital design centered only on technological or economic innovation, and above all the emergence of other paths [GRA 13] of digital technologies and interaction design, in the service of the common good. Social and digital innovation thus appears as "a type of collaborative innovation in which innovators, users and communities collaborate with the help of digital technologies in order to co-create knowledge and solutions that respond to a wide range of social needs"[10]. Thus, social and digital innovation is an apt approach for rethinking informational ecology through design in the field of health.

From this perspective, we can envisage the development of an interdisciplinary research program, at the intersection of humanities and social sciences, medical science, engineering sciences and design sciences to contribute to the field of reflection and conception of m-health. The goal of this program would be double: on the one hand it would involve studying the use (and misuse) of connected objects and specialized applications in the field of m-health and, on the other hand, to use project research to contribute to the development of digital devices that integrate an ethical and social dimension from their conception. Starting from a logic of co-design between the different stakeholders in the sector, it would involve initiating a collaborative design process between doctors, patients, public and private institutions. The problem of digital design joins that of public policy design once the devices studied and produced come from a public health policy that needs to be questioned in order to be updated. On the part of users, this program would also allow the development of therapeutic education for patients through digital and human mediation, on the one hand by observing how digital interfaces and connected objects transform the doctor/patient

---

10 This definition of social and digital innovation was offered by the researchers at the European Digital Social Innovation Project in 2014: http://content.digitalsocial.eu/about/.

relationship, and on the other by offering new digital tools, based on a logic of human-centered design. As a complement to the observational study on the use of connected objects and digital interfaces in the field of health, the interventional study would have the goal of testing a digital device to serve the needs of stakeholders, which are the doctors and patients, which could contribute to the search for solutions. The research program would include three phases of research (qualitative research, conception, implementation), by articulating theory and empirical data, effective conception and production of a digital device *by* and *for* the beneficiaries of the m-health project. This type of project research would finally make it possible to place the role of digital medical devices within a more global system of education and prevention, by replacing the human at the center of its concerns.

6.5. Conclusion

To conclude, there is an ambiguity maintained by promotional discourse between m-health and Quantified Self. Portable information technologies used in the Quantified Self fall within the continuity of this history of information technologies and incorporate anthropotechnics. To respond to the problem, we can say that connected objects and the applications of the Quantified Self transform the relationship to body and to its representation. The body becomes in effect quantified and quantifiable, transparent and readable. Thus, the Quantified Self and the Internet of Things transform the body into a networked resource. From a critical perspective, we can ask if the body and all of existence can be reduced to a sum of data and "information behaviors" [PUC 14].

Analysis of Chris Dancy's use of communicating objects has made it possible to comprehend the dominant ideology in the field of information technologies and interaction design. This ideology draws on an image of technologies oriented toward transhumanism and emotional interaction design. It is also part of the legacy of the ubiquitous computing project developed by the beginning of the 1990s in the United States. It also anticipates the incorporation of technologies by drawing on an imagery tinged with animism and anthropomorphism, and a symbiotic relationship between interfaces, humans and their environment.

This immersion in our private lives through our behavioral and computer data must be placed within the "full vision" [WAJ 10] created by the multiplication of screens, symptomatic of our era's zeitgeist. We can critically consider the future of Big Data in this quest for total transparency:

"do we consider this technological path as a social evolution? Is it the dream of every citizen to transform his body into a connected object, into a statistical body, network resource available as open-data, or even into an API?" The extreme experience of controlling his own data recently led Chris Dancy to have a real identity crisis: he has been "devoured by his own data" [RIC 16] as he explained it in his recent interviews.

Furthermore, beyond analysis of Chris Dancy's case, we have raised, in the last part of the article, the ethical and social challenges related to the governance of data. This entails rethinking m-health considering social support, a distinctive feature of the French healthcare system.

We have also sought to explain the necessity of developing, by completing a critical analysis of dominant technologies in the m-health sector, a form of participative research, falling along the lines of project research [FIN 15]. It would make it possible to address the current problems of doctors and patients, whose relationship is transformed by the surge in connected objects, in order to offer alternative solutions.

Other *paths* [GRA 13] remain to be imagined in the field of portable information technologies specializing in health. An interesting pathway to explore would be that of "clearing desirable paths for transforming "humanitude" and the mechanisms of action that would lead there (political, social, educational)" [GOF 13]. Rethinking mobile applications and connected objects in a design logic centered on the human and oriented toward social innovation seems to be a pertinent direction for research. Research in humanities and social sciences can thus contribute to the development of socially innovative digital devices, by taking into account the needs of users, considered to be stakeholders in a public health project. From a socio-political point of view, project research in design and m-health is a participative research method that allows citizens to get involved in the current public debate about the management of personal health data, in order to contemplate solutions centered on patient interests and in dialogue with healthcare professionals.

6.6. Bibliography

[BER 67] BERTIN J., *Sémiologie graphique: diagrammes, réseaux, cartes*, Mouton, Paris, 1967.

[BEY 12] BEYAERT-GESLIN A., *Sémiotique du design*, PUF, Paris, 2012.

[BON 15] BONENFANT M., PERRATON C., *Identité et multiplicité en ligne*, Presses Universitaires du Québec, Montreal, 2015.

[BOU 02] BOULLIER D., "Objets communicants, avez-vous donc une âme?", *Les Cahiers du numérique*, vol. 3, no. 4, pp. 45–60, 2002.

[BUC 01] BUCHANAN R., "Human Dignity and Human Rights: Thoughts on the Principles of Human-Centered Design", *Design Issues*, vol. 15, no. 3, p. 35, 2001.

[CAR 05] CARDON D., *A quoi rêvent les algorithmes?*, Le Seuil, Paris, 2005.

[CAT 16] CATOIR-BRISSON M.-J., CACCAMO E., "Métamorphoses des écrans: invisibilisations", *Interfaces numériques*, vol. 5, no. June 2, 2016.

[CAT 12] CATOIR M.-J., LANCIEN T., "Multiplication des écrans et relations aux médias: de l'écran d'ordinateur à celui du smartphone", *MEI*, no. 34, pp. 53–65, 2012.

[CEN 14] CENTRE D'ETUDES SUR LES SUPPORTS DE L'INFORMATION MEDICALE, Baromètre annuel sur les usages digitaux des professionnels de santé, CESSIM-Ipsos, 2014.

[CLA 14] CLAVERIE B., "De la cybernétique aux NBIC: l'information et les machines vers le dépassement de l'human", *Hermès La Revue*, no. 68, pp. 95–101, 2014.

[COM 09] COMYN G., "La e-santé: une solution pour les systèmes de santé européens?", *Les dossiers européens*, no. 17, May–June 2009.

[COU 15] COUTANT A., STENGER T. (eds.), *Identités numériques*, L'Harmattan, Paris, 2015.

[CNI 14] CNIL, "Quantified-Self, m-santé: le corps est-il un nouvel connecté?", CNIL, available at: http://www.cnil.fr/linstitution/actualite/article/article/quantified-self-m-sante-le-corps-est-il-un-nouvel-objects-connecte/, 2014.

[CON 15] CONSEIL NATIONAL DE L'ORDRE DES MEDECINS, Santé connectée: de la e-santé à la santé conectée, CNOM White Paper, January 2015.

[CUK 14] CUKIER K., MAYER-SCHÖNBERGER V., *Big data: la revolution des données est en marche*, Robert Laffont, Paris, 2014.

[DEM 02] DEMASSIEUX N., "Au-delà de la 3G, les communicating objects", *Les Cahiers du numérique*, vol. 3, no. 4, pp. 15–22, 2002.

[DEN 05] DENI M., "Les objects factitifs", in FONTANILE J., ZINA A. (eds), *Les objects au quotidien*, Presses Universitaires des Limoges, 2005.

[FIN 15] FINDELI A., "La recherche-projet en design and la question de recherche: essai de clarification conceptuelle", *Sciences du Design*, no. 1, pp. 43–55, 2015.

[FLO 90] FLOCH J.-M., *Sémiotique, marketing and communication*, PUF, Paris, 1990.

[GAU 00] GAUDREAULT A., MARION P., "Un média naît toujours deux fois…", *Sociétés et Representations*, no. 9, pp. 21–36, 2000.

[GOF 13] GOFFETTE J., "De l'humain à l'humain augmenté: naissance de l'anthopotechnics", in KLEINPETER E. (ed.), *L'humain augmenté*, CNRS Editions, Paris, 2013.

[GOF 06] GOFFETTE J., *Naissance de l'anthopotechnie: De la biomédecine au modelage de l'human*, Vrin, Paris, 2006.

[GRA 13] GRAS A., *Les imaginaires de l'innovation technique*, Manucius, Paris, 2013.

[JAR 14] JARRIGE F., *Technocritiques: du refus des machines à la contestation des technosciences*, La Découverte, Paris, 2014.

[LAM 14] LAMONTAGNE D., "La culture du moi quantifié – le corps comme source de données", ThotCursus, available at: http://cursus.edu/article/22099/culture-moi-quantifie-corps-comme-source/#.U_dcR0sQ4b8, 2014.

[LAN 10] LANCIEN T., "Multiplication des écrans, images et postures spectatorielles", in BEYLOT P., LE CORFF I., MARIE M. (eds), *Les images en question, Cinéma, télévision, nouvelles images*, PUB, Bordeaux, 2010.

[MAN 07] MANZINI E., "Design Research for Sustainable Social Innovation", in MICHEL R. (ed.), *Design Research Now: Essays and Selected Projects*, Birkhäuser, Basel, 2007.

[MAN 15] MANZINI E., *Design, When Everybody Designs, an Introduction to Design for Social Innovation*, MIT Press, Cambridge, 2015.

[MER 13] MERZEAU L., "L'intelligence des traces", *Intellectica*, no. 59, pp. 115–135, 2013.

[MIC 15] MICHAL-TEITELBAUM C., "Big Data et Big Brother. Données et secret médical, vente de dossiers médicaux aux sociétés privés et médecine personnalisée", available at: http://docteurdu16.blogspot.fr/2015/04/Bigdata-et-big-brother-donnees-et.html, 2015.

[MOR 62] MORIN E., *L'esprit du temps*, Grasset-Fasquelle, Paris, 1962.

[MPO 13] M'PONDO DICKA P., "Sémiotique, numérique et communication", *RFSIC*, no. 3, 2013.

[NOR 12] NORMAN D., *Design émotionnel*, De Boeck, Brussels, 2012.

[OMS 11] ORGANIZATION MONDIALE DE LA SANTÉ, "Health New Horizons for Health Through Mobile Technologies", *Global Observatory for eHealth series*, vol. 3, 2011.

[OMS 46] ORGANIZATION MONDIALE DE LA SANTE, "Définition de la santé, Préambule à la constitution de l'OMS" *Conférence internationale sur la Santé*, New York, United States, June 19–22, 1946.

[PAP 85] PAPANEK V., *Design for the Real World: Human Ecology and Social Change*, Academy Chicago Publishers, Chicago, 1985.

[PIN 02] PINTE J.-P., "Introduction", *Les Cahiers du numérique*, vol. 3, no. 4, pp. 9–14, 2002.

[PRI 02] PRIVAT G., "Des objets communicants à la communication ambiante", *Les Cahiers du numérique*, vol. 3, no. 4, pp. 23–44, 2002.

[PUC 14] PUCHEU D., "L'altérité à l'épreuve de l'ubiquité informationnelle", *Hermès La revue*, no. 68, pp. 115–122, 2014.

[REN 14a] RENUCCI F., "L'homme interfacé, entre continuité et discontinuité", *Hermès La revue*, no. 68, pp. 203–211, 2014.

[REN 14b] RENUCCI F., LE BLANC B., LEPASTIER S., "L'autre n'est pas une donnée. Altérités, corps and artefacts", *Hermès La revue*, no. 68, 2014.

[RIC 16] RICHARD C., "L'homme le plus connecté du monde s'est fait dévorer par ses données", L'Obs-Rue 89, http://rue89.nouvelobs.com/2016/06/17/ lhomme-plus-connecte-monde-sest-fait-devorer-donnees-264377, June 17, 2016.

[ROU 14] ROUVROY A., "Avant-Propos - Du Quantified-Self à la m-santé: les new territoires de la mise en données du monde", *Cahiers IP de la CNIL: Le corps, nouvel object connecté*, no. 2, pp. 4–5, 2014.

[ROU 13] ROUVROY A., BERS T., "Gouvernementalité algorithmique et perspective d'émancipation", *Réseaux*, no. 177, 2013.

[SAL 08] SALMON C., *Storytelling, la machine à fabriquer des histoires et à formater les esprits*, La Découverte, Paris, 2008.

[SOO 11] SOOJUNG-KIM P., "Contemplative Computing", *Conférence du Microsoft Research*, available at: https://www.academia.edu/635387/Contemplative_Computing, Cambridge, United States, 2011.

[SUC 11] SUCHMAN L., "Anthropological Relocations and the Limits of Design", *Annual Review of Anthropology*, no. 40, pp. 1–18, 2011.

[TRE 14] TRELEANI M., "Bientôt la fin de l'écran", *INA Global*, no. 1, pp. 64–70, 2014.

[VIA 15] VIAL S., *Le design*, PUF, Paris, 2015.

[VID 13] OBSERVATOIRE VIDAL, Usages numériques en santé, Deuxième baromètre sur les médécins utilisareurs de smartphone en France, May 2013.

[WAJ 10] WAJCMAN G., *L'œil absolu*, Denoël, Paris, 2010.

[WEI 91] WEISER M., "The Computer for the 21st Century", *Scientific American*, vol. 265, no. 3, pp. 66–75, 1991.

Tweets from Fukushima: Connected Sensors and Social Media for Dissemination after a Nuclear Accident

7.1. Introduction

Digital services for information and communication, especially social media, are increasingly being used to share the information required to manage disasters such as a nuclear accident. In this particular case, victims can only rely on measurements to evaluate the radioactive contamination of the environment, food and people. Measurement readings are therefore essential in implementing actions towards the reduction of people's exposure to ionizing radiation and the monitoring of its health impact. Thus, in a situation post nuclear accident, it is crucial to have access to measurement devices as well as tools facilitating the dissemination of information useful to the resilience process.

In March 2011, a few days after the Fukushima Daiichi nuclear accident, Japanese citizens sought to obtain information about the radioactive contamination of their environment. Without complete information from public institutions they extensively relied on social media to find the level of radioactive contamination in different areas and assess practical solutions to ensure their everyday life. This data was partially generated by *ad hoc* devices, called radiation detectors or radiameters, and disseminated on the Internet by automated programs (robots) through the Twitter platform. This way, radiameters contributed to the Internet of Things (IoT).

Chapter written by Antonin SEGAULT, Frederico TAJARIOL and Ioan ROXIN.

In order to study their role, we have conducted a study on the dissemination of information via social media after a nuclear accident. Our work is part of a research project on the use of social media in post nuclear accident situations, SCOPANUM (*Stratégies de communication de crise en gestion post accident nucléaire via les médias sociaux* or Post-Nuclear Accident Crisis Management Communication Strategies via Social Media)[1].

This chapter is structured as follows: after introducing the IoT (section 7.2) and reviewing the essential elements of the role of social media in a crisis situation (section 7.3), we will describe the context, method and results of our study (sections 7.4 to 7.9).

7.2. The IoT: a shift in the development of digital services

The number of objects connected to the Internet is now larger than the number of people who use the Internet to communicate. In 2003, the human population was over 6.3 billion people[2] and about 500 million objects were connected to the Internet[3], the equivalent of less than one device per individual. By 2010, when the population numbered around 6.8 billion people, the Internet ecosystem contained more than 12.5 billion connected objects[4], which is roughly two devices per individual. Predictions for 2020 suggest that 7.6 billion people will be sharing around 40 billion connected objects[5], a ratio of between five and six connected objects per person. This ratio might even be higher, since in 2015 more than half of the population (57%) did not have any regular access to the Internet[6].

---

1 http://semlearn.pu-pm.univ-fcomte.fr/scopanum.

2 http://www.census.gov/population/international/data/worldpop/table_population.php.

3 "Forrester CEO Forecasts Web Services Storm" (2003), consulted 2.26.2016: http://www.computerweekly.com/news/2240049850/Forrester-CEO-forecasts-web-services-storm.

4 World Internet Stat, Usage and Population Statistics. http://www.Internetworldstats.com/stats.htm.

5 Sorrell S. (2015), "The Internet of Things. Report of Juniper Research" consulted February 26, 2016:http://www.juniperresearch.com/document-library/white-papers/iot-Internet-of-transformation.

6 The State of Broadband 2015: Broadband as a Foundation for Sustainable Development. Report by the broadband commission for digital development. International Telecommunication Union (ITU) and United Nations Educational, Scientific and Cultural Organization (UNESCO).

These numbers indicate significant change in the development of a specific type of Internet service, known as the Internet of Things (IoT)[7]. These services exchange data between objects connected to the Internet and enable the diffusion of this data to users of these devices. This shift is not the result of technological breakthroughs, but rather the convergence of many factors: the spread of mobile devices (smartphones, tablets), the improvement of bandwidth, expansion of wireless connections, as well as the integration of micro-electronic components into physical objects of smaller dimensions and their capacity to generate and transfer data [VER 14].

Each connected object, with its user identification, is an integral part of the Internet [XIA 12], thus extending the idea of computational ubiquity [WEI 91]. In this context, APIs (Application Programming Interfaces) allow different digital services to freely access subsets of data made available by other services, and thus to mix heterogeneous data (mash-up) and construct new useful representations for the user [END 13].

Thanks to the pervasiveness of mobile devices and physical objects equipped with sensors connected to the Internet, the digital services of the IoT make it possible to control a situation in real time, provide users with enriched information and access other Internet services, such as social media, at any time.

The objects of the IoT are numerous and are related to different aspects of human life, such as home automation (when sensors installed in a weather station send a signal to lower the shades of all the houses in a geographical area [END 13]), or computer-assisted driving of a car (where the onboard navigator provides alternative routes in real time, taking into account traffic data sent by other connected vehicles [ZAN 14]).

Our study deals with a type of connected object, the radiameter, which is a physical object endowed with radioactivity sensors (Figure 7.1).

---

7 Although the concept had already been used by [GER 99], the term IoT appeared for the first time in an article in *Forbes* Magazine (SCHOENBERGER C.R., "The Internet of Things", March 2002) and was popularized by Kevin Ashton (Auto-ID center, Massachusetts Institute of Technology). It spread further in 2005 with the publication of a volume by the International Telecommunication Union [INT 05] and then the first international conference on the subject [FLO 08].

Figure 7.1. *A connected radiation detector (Poket Geiger™ type 4)*

The radiation detector belongs to the category of objects intended for the protection of citizens [ROS 14] and, once connected to the Internet, it can spread useful information for the survivors of a nuclear accident.

7.3. Social media and the dissemination of information during a catastrophe

Social media services are digital services available on the Internet (for example, YouTube, Facebook, Twitter, etc.) that allow users to spread information, freely and on a large scale, through the publication, retransmission and sharing of content. Thanks to these services, any user is free to create public or private profiles, to manage lists of connections with the profiles of other users, and to navigate through these lists [BOY 08]. Consequently, every social media user becomes simultaneously a producer, broadcaster and consumer of information [BRU 08] in real time thanks to mobile supports. For example, in 2013, 75% of Twitter users accessed their Twitter accounts via a mobile support [LUN 13].

On Twitter, the diffusion of information is ensured not only by real users but also by automated programs, called robots or bots, which produce a large proportion of the messages [CHU 10]. By means of social media APIs (for example API-Twitter), the data published by these robots can then be processed by other programs or connected objects (Figure 7.2).

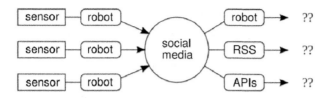

Figure 7.2. *Social media and the Internet of Things*

Some of these robots are developed with a malicious intent (such as for spamming or phishing), but others offer functions for processing, aggregating or rebroadcasting interesting content and can also be used to automatically share data produced by connected sensors on social media. For example, in 2008, artists of the Botanicalls[8] collective developed an IoT system in the form of humidity sensors that send a message when a plant needs to be watered. Likewise, for several years the *@twrbrdg_itself*[9] Twitter account (created by designer Tom Armitage) published messages indicating the opening and closing of Tower Bridge to help London drivers (Figure 7.3).

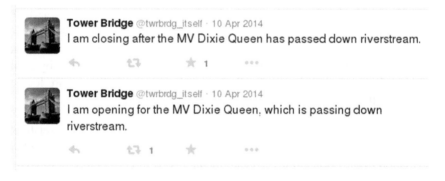

Figure 7.3. *Tweets from the @twrbrdg_itself account*

The convergence between social media, robots and connected objects can also be seen as a channel of communication contributing to the IoT [KRA 10] from a participative perspective [ORE 05]. This participatory attitude is a useful lever after a disaster to share information and allow communication between people.

---

8 http://www.botanicalls.com/.

9 http://twitter.com/twrbrdg_itself.

Over the past few years, people facing crisis situations have frequently turned to social media to share crucial information and coordinate their actions [PAL 10]. For example, during the Virginia Tech shooting in 2007, students and their relatives were able to collaboratively and precisely identify the victims using Facebook [PAL 09]. During the forest fires and floods in the United States in 2009, Twitter users shared a wide variety of information to help evaluate the situation (for example: weather, state of the roads, advice, requests for help, geographical information, etc.) [VIE 10]. Certain social media platforms have moreover set up specific tools, such as Facebook's Safety Check[10] or Google's Person Finder[11], as a way to help users get information about their loved ones during disasters.

Specific uses have emerged *during* and *after* natural or industrial disasters in order to facilitate the diffusion, search and corroboration of information. The analysis of messages spread over Twitter during certain earthquakes (for example in New Zealand, Italy, etc.) has shown that Twitter users have adopted hashtags to facilitate the diffusion of information during aftershocks [BRU 12b] and that the automatic spread of information about the earthquake responded to precise needs on the part of citizens [COM 15].

Furthermore, victims favor the rebroadcast of information and the publication of hypertext links rather than the exchange of personal messages [HUG 09]. Thanks to these collaborative practices, the crisis situation can benefit from management that is both highly parallel – carried out by many people at the same time – and distributed – involving people who are located in different places, even throughout the world [PAL 10].

Studies on the dissemination of information via social media are relatively recent. Despite the generic recommendations for how institutions can deliver information to the population via social media [WHI 09], it is difficult to predict their use in each crisis situation. We have thus chosen to illuminate another possible crisis situation, the one following a nuclear accident.

7.4. Context of the study

We conducted a study on the diffusion of information via social media during the post-accident phase of a radiological accident. The management of a nuclear accident is, according to convention, divided into two

---

10 http://www.facebook.com/about/safetycheck/.

11 http://google.org/personfinder/global/home.html.

temporally distinct phases: the emergency phase, during which radioactive substances are released into the environment, and the post-accident phase, during which the consequences of the accident must be managed (Figure 7.4). This second phase is itself generally divided into two successive periods: the transition period, when the contamination of the area is still not completely known, then the long-term period [COD 12].

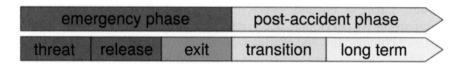

Figure 7.4. *Phases of a radiological accident*

During this second period, which can last several decades, people living in the contaminated areas undergo prolonged exposure to low doses of radiation. Since radiation cannot be perceived by human senses, only measurements make it possible to evaluate the radiological situation. The production and the sharing of these measurements are therefore decisive for the choice, implementation and monitoring of solutions intended to limit the population's exposure. Depending on the level of radiation measured, these solutions are distributed along a continuum that goes from food restrictions up to the evacuation of locations. Ambient radioactivity, resulting from the contamination of the environment by radionuclides, is usually measured with the help of a radiation detector or radiameter, also called a Geiger counter, after the fact that the sensor most frequently used by radiation detectors is the Geiger-Müller tube. Thanks to these measurement readings, survivors could decide to leave or stay to reconstruct, and thus begin the process of resilience, which consists of reestablishing a material, psychological, personal and societal equilibrium [CUT 13].

The context of our study is the accident that took place at the Japanese Fukushima Daiichi Nuclear Power Plant in March 2011. The failures caused by the earthquake and tsunami of March 11, 2011, led to the release of large quantities of radioactive particles into the environment. This accident, assessed at level seven (the highest) on the International *Nuclear Event Scale*[12] (INES), required the evacuation of more than 100,000 people [GOR 14].

---

12 http://www.irsn.fr/FR/connaissances/Installations_nucleaires/La_surete_Nucleaire/echelle-ines/.

Immediately after the accident, the population only had access to limited information regarding measurements of radioactive leaks and its potential impact on health [ALD 12]. However, after a few days this information was revealed to be incomplete and the population had little confidence in the information provided by the Japanese government and by TEPCO, the company that owned the Fukushima Daiichi Nuclear Power Plant [LI 14].

Thus, Japanese citizens began to carry out their own radioactivity measurements by developing accessible measuring devices, building mutual assistance communities and sharing (for example Safecast[13], Pokega[14]) [KER 13]. These citizens also relied on social media to collect scarce data distributed by the Japanese authorities, aggregating it with measurements carried out by the citizens themselves, and thus created cartographic representations of the contamination of the area [PLA 11] (Figure 7.5)[15].

Figure 7.5. *Collaborative chart of the contamination*

International and Japanese radioactivity experts and laypeople used social media to share their knowledge about the effects and the dynamics of radioactive measurements, to comment on the accuracy of the data provided by the authorities in charge of the crisis and to discuss the pertinence of the information sources [FRI 11]. Thanks to social media, citizens could popularize the specialized language of experts and understand how to deal with the situation.

---

13 http://blog.safecast.org/.

14 http://www.radiation-watch.org/.

15 http://japan.failedrobot.com.

7.5. Goals of our study

In this context, we were interested in the forms and processes of information dissemination via social media during the post-accident phase. Our study focuses on a group of automated programs – or bots – created on Twitter between 2010 and 2014. These programs use user accounts on the microblogging platform to share, in real time, measurements of ambient radioactivity carried out by connected radiation detectors. These measurements are published at regular intervals in the form of tweets, which are short text messages with a maximum of 140 characters. These tweets can also contain hypertext links or hashtags (which are keywords marked by a # symbol and used to annotate tweets by creating folksonomies [PET 11]). Each user profile connected to a robot contains metadata describing the measurement system (Figure 7.6).

Figure 7.6. *A bot's user profile*[16]

7.6. Methodology

We chose to study the Twitter platform for several reasons. First of all, research on the use of social media in disaster situations has shown the dominant role of Twitter [ASH 14]. In fact, this platform provides a very simple structure, notably enabling unilateral connections between users, while others, such as Facebook, require reciprocity [BRU 12a]. Thus, a Twitter user can receive tweets from any other user via the subscription system (following), while on Facebook the user must accept another user and also be accepted. Moreover, from the methodological point of view,

16 http://twitter.com/NeduMP.

unlike other forms of social media, Twitter provides access to a large section of the *tracks* of users thanks to programming interfaces (API).

The concept of "information diffusion", although discussed several times in the literature [GOM 10, GRU 04, LER 10, LIB 08], does not provide metrics for our study. For this reason, we have operationalized the concept of diffusion according to three dimensions – the popularity of the robots, the completeness of the measurements that they share and the source of these measurements. The popularity of robots is examined with the help of quantitative metrics for Twitter accounts: the number of followers (which means the users who subscribed to the account's publications), listed (i.e. the inclusion of the account in thematic lists), retweets (i.e. the redistribution of messages by other users) and favorites (i.e. saving the messages sent by other users). To evaluate the completeness of the measurements, we focused on the unit of measurement, the precision, the type of device and the place where the measurement was carried out in each tweet's content and in the bot's profile. Finally, regarding the source of the measurements shared, we verified if the Twitter accounts were publishing original measurements – that is to say those coming from a measurement tool managed by the bot's creator – or were just rebroadcasting the measurements produced by other sources.

We first created a list of Twitter accounts which automatically share radiation measurements. To do this, first of all we collected, through the Search API[17], the latest tweets containing key words relating to the units of measurement for radiation: 'cpm'; 'gy/h'; 'µgy'; 'ngy'; 'usv'; 'µsv'; 'sv/h'. Among the user accounts that produced these tweets, we were able to identify 48 active robots, of combining a search of regular expressions and manual sorting. We then used Twitter's RESET API[18] and Streaming API[19] to collect a large amount of data from the bots' user profiles, as well as their 1,000 most recent tweets, used in particular to calculate the number of *retweets* and *favorites*.

7.7. Results

First, we present general results on the corpus of bots, then we detail the analysis of the spread according to three dimensions: popularity of the bots, comprehensiveness and source of measurements:

---

17 http://dev.twitter.com/rest/public/search.

18 http://dev.twitter.com/rest/public.

19 http://dev.twitter.com/streaming/overview.

7.7.1. *Comprehensive overview*

The bots that we identified (N=48) were created from year of the disaster, 2011 (40%, 15% of which were in the month of March), then in 2012 (30%) and in 2013 (25%). The most common language specified in the profile is Japanese (88%), the time zone is that of Tokyo (42%), and the tweets contain Japanese characters (75%). These elements suggest that the information was shared by and meant for users fluent in Japanese.

The frequency of the publication of tweets varies from 10 minutes to 12 hours, but mostly falls between 30 minutes (44%) and an hour (31%). Moreover, certain Twitter accounts are also used to share data about values from other types of sensors (29%) such as thermometers or anemometers[20] or non-automated tweets. Finally, 60% of the bots used *hashtags*, the most common of which, #Radidas and #Mark2bot, refer to the devices that we detail in the following sections. The other hashtags referred to radioactivity (#geiger, #jp_geiger, #genpatsu[21]) or places (#musashino, #ibaraki, #yokohama).

7.7.2. *Popularity of bots*

An analysis of the data shows (Figure 7.7) that the popularity, in terms of number of *followers*, *listed*, *retweets* and *favorites*, follows a long-tail distribution: only a few of the bots are very popular (N=5 with more than 1,000 *followers*), while the majority are only slightly popular, with a median value of 23 followers.

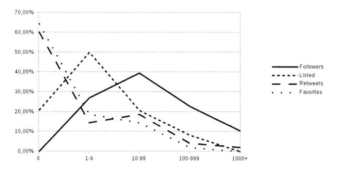

Figure 7.7. *Distribution of the popularity of bots*

---

20 Measuring the temperature and the speed of the wind respectively.
21 This means "nuclear power plant" in Japanese.

Moreover, the bots created just before and immediately after the Fukushima Daiichi accident (N=2 in January 2011, N=7 in March 2011) are among the most popular, while the bots created later are considerably less popular, and their popularity decreases progressively with time (Figure 7.8).

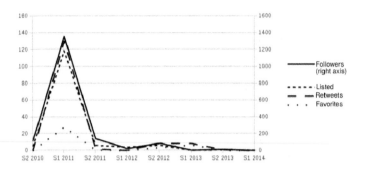

Figure 7.8. *Popularity (average) of bots according to their date of creation (per semester)*

7.7.3. *Completeness of the shared measurements*

All of the bots share the value of the radiation measurements by indicating the unit of measurement used. The unit of measurement used most often is Sieverts per hour (58%) and more rarely Grays per hour (13%). Several bots (21%) share measurements in two different units, particularly in Sieverts per hour and in Counts per Minute. Only one bot uses Röntgen, an obsolete unit of measurement[22].

The precision of the measurements broadcast varies strongly among the bots. In Sieverts per hour, the degree of precision is mainly 1 nSv/h (49%) which is 0.01 µSv/h (44%). In Grays per hour, the most frequent precision is 1 nGy/h (71%). In Counts per Minute, it is 0.1 CPM (67%). However, these statistics don't take into account the mathematical precision of the values presented in the tweets, which could differ from the resolution of the sensors used. Only 23% of the bots (those using the Radidas system, described further on) explicitly indicate their precision (example: "±0.01µSv/h").

Significant differences can also be seen in the location displayed by the bots. The geographical coordinates of the measuring point are not provided

22 http://www.nist.gov/pml/pubs/sp811/sec05.cfm.

by 33% of the bots, while the other bots provide less precise data, such as the name of the town (31%) or the neighborhood. A minority of bots (10%) give no indication of where the measurement was taken.

7.7.4. *Source of the measurements shared*

The source of the measurements is divided between original or rebroadcast. Almost half (44%) of the bots do not specify the source of the measurements that they share, and 17% of the bots explicitly indicate rebroadcasting of data produced by an official measurement organization. The other bots (40%) spread original information, coming from a connected radiation detector. Among the 48 bots, only 25% provide the name of the measuring device used, the others are satisfied with more vague data such as a photograph or a description of the device.

The hashtag*s* #Radidas and #mark2bot have allowed us to identify two devices enabling the publication of radiation measurements, which we call "ready-to-use robots" since they are easy to implement and don't require advanced knowledge in computer science or electronics: Radidas and Mark2.2. Radidas, used by 27% of the bots, is a software program which enables the sharing of data generated by a radiation detector connected to a computer (Figure 7.9). The radiation detector Mark2, used by 8% of the bots, is designed to automatically publish measurements on Twitter.

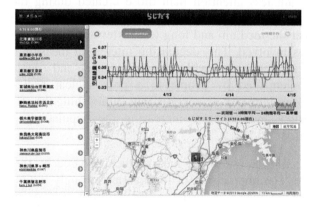

Figure 7.9. *Screen capture of the Radidas software*[23]

---

23 http://pow2p.web.fc2.com/pgnet/sample/.

7.8. Discussions

Social media is an important lever for the diffusion of information, especially during natural or industrial disasters. We have studied the spread of measurements from connected sensors via social media during the post-accident phase of the Fukushima Daiichi nuclear accident. We were interested in the forms and processes of the dissemination of information relating to radiation measurements via Twitter. We have identified N=48 bots that automatically broadcast messages containing measurements of ambient radioactivity. We have analyzed the content of these tweets, as well as the robots' user profiles. To study the spread of tweets from bots, we have considered three dimensions: popularity of the bots, the completeness of the measurements that they share and the source of these measurements.

The analysis of the nature of the bots (language, date of creation, number of accounts following) seems to confirm that they have performed a support function for the information dissemination in the crisis situation after the Fukushima Daiichi nuclear accident of 2011. Furthermore, since a large portion of the bots diffuse original measurements, we regard these tools as falling within the same collaborative practices as those already identified around the aggregation and mapping of measurements [PLA 11].

The popularity of the accounts rapidly decreased, even though they were very important in the months following the accident. The bots created a few months later gain less attention than those created during the emergency phase, whatever the type of source. This could be an indicator of a decrease in interest, over the medium and long term after the accident, on the part of the populations involved. This decreasing interest in measurement and risk management is consistent with the avoidance and denial strategies sometimes implemented by people in a post nuclear accident situation [VAN 90]: diminishing risk perception, thus diminishing the effectiveness of counter measures and therefore increasing the population's exposure to ionizing radiation.

Finally, incompleteness of the measurements shared by the bots affects their reliability. The imprecision of localization strongly limits the value of the data, contamination of an area can very strongly vary over distances of several meters. The lack of information on the type of device used also constitutes an obstacle when comparing data. The impossibility of aggregating the measurements from different sources and different bots was an impediment to their verification, to the search for a consensus and the detection of possible errors. These weaknesses reduce the quality of the information available to the users of Twitter bots, and threatening to their decision-making and the success of the resilience process.

7.9. Conclusions

The supply of digital services dedicated to information and communication is expanding due to the connected objects that make up the IoT. All parts of human life are involved, including situations resulting from a natural or industrial disaster. We have presented a study on a specific connected object the radiation detector, capable of measuring radioactivity levels of an area, thanks to sensors, and share them via the Internet on Twitter or other platforms. This configuration may thus provide useful measurements for the survivors of a nuclear accident and ensure these measurements are shared.

The results of our research open up avenues for reflection regarding the design of communicating objects, such as Twitter bots, adapted to a post-nuclear accident situation. First, the decrease in popularity of the bots underlines the necessity of supporting the involvement of citizens in carrying out and broadcasting measurements. To this end, two directions seem useful to us. First, it seems pertinent to us to increase the sharing of measurements made by existing devices, for example by setting up aggregation tools such as interactive maps. Next, it is necessary to reduce the obstacles with which citizens can be confronted, by wider distribution of the basic knowledge and technical skills necessary for understanding the radiation measurements shared by the robots as well as for the setup of new connected radiation detectors.

Moreover, the completeness of the measurements can only improve the reliability and utility of the data.

The results of the two investigations, conducted by a network of experts in radiation protection and communities of amateurs, show that the pairing "radiation detector and Twitter bots" can contribute to ensuring the completeness of the measurements, provided that the metadata is established. Our research has made it possible to identify the most useful metadata, as much for experts as for non-experts, and thus come up with recommendations meant for the creators of the bots.

Later work, aimed more specifically at the study of user profiles, is necessary to establish predictive and descriptive models of the trajectories and the flow of information between the creators and followers of the bots. The initial conclusions of this work will nevertheless be used, as part of the SCOPANUM Project, for the creation of digital services to help spread information within populations who live in the contaminated zone following a nuclear accident and have access to a Pocket Geiger™ radiation detector.

7.10. Acknowledgements

This work was carried out as a part of the SCOPANUM project, a partnership between CEPN[24] and the OUN team[25] at the ELLIADD Laboratory[26], financed by the *Conseil supérieur de la formation et de la recherche stratégique*[27], and recipient of a doctoral contract funded by the Pays de Montbéliard township committee[28].

7.11. Bibliography

[ALD 12] ALDRICH D.P., "Post-crisis Japanese Nuclear Policy: From Top-down Directives to Bottom-up Activism", *Asia Pacific Issues*, vol. 103, no. 1, pp. 1–12, 2012.

[ASH 14] ASHKTORAB Z., BROWN C., NANDI M. *et al.*, "Tweedr: Mining Twitter to Inform Disaster Response", *Proceedings of the 11th International ISCRAM Conference*, Penn State, United States, 2014.

[BOY 08] BOYD D.M., ELLISON N.B., "Social Network Sites: Definition, History, and Scholarship", *Journal of Computer-Mediated Communication*, vol. 13, no. 1, pp. 210–230, 2008.

[BRU 12a] BRUNS A., "How Long is a Tweet? Mapping Dynamic Conversation Networks on Twitter using Gawk and Gephi", *Information, Communication & Society*, vol. 15, no. 9, pp. 1323–1351, 2012.

[BRU 12b] BRUNS A., BURGESS J.E., "Local and Global Responses to Disaster: #eqnz and the Christchurch Earthquake", *Proceedings of the Disaster and Emergency Management Conference*, Brisbane, Australia, pp. 86–103, 2012.

[BRU 08] BRUNS A., *Blogs, Wikipedia, Second Life, and Beyond: From Production to Produsage*, Peter Lang, New York, 2008.

---

24 http://www.cepn.asso.fr.

25 http://semlearn.pu-pm.univ-fcomte.fr.

26 http://elliadd.univ-fcomte.fr.

27 http://csfrs.fr.

28 http://www.agglo-montbeliard.fr.

[CHU 12] CHU Z., GIANVECCHIO S., WANG H. *et al.*, "Who is Tweeting on Twitter: Human, Bot, or Cyborg?", *Proceedings of the 26th Annual Computer Security Applications Conference (ACSAC '10)*, Austin, United States, pp. 21–30, 2010.

[COD 12] CODIRPA, "Eléments de doctrine pour la gestion post-accidentelle d'un accident nucléaire", available at: http://post-accidentel.asn.fr/content/ download/53098/365511/version/1/file/Doctrine_CODIRPA_NOV2012.pdf, 2012.

[COM 15] COMUNELLO F., MULARGIA S., POLIDORO P. *et al.*, "No Misunderstandings During Earthquakes: Elaborating and Testing a Standardized Tweet Structure for Automatic Earthquake Detection Information", *Proceedings of the 12th International Conference on Information Systems for Crisis Response and Management (ISCRAM 2015)*, Kristiansand, Norway, 2015.

[CUT 13] CUTTER S.L., AHEARN J.A., AMADEI B. *et al.*, "Disaster Resilience: A National Imperative", *Environment: Science and Policy for Sustainable Development*, vol. 55, no. 2, pp. 25–29, 2013.

[END 13] ENDRES-NIGGEMEYER B. (ed.), *Semantic Mashups. Intelligent Reuse of Web Resources*, Springer-Verlag, Berlin, 2013.

[FLO 08] FLOERKEMEIER C., LANGHEINRICH M., FLEISCH E. *et al.* (eds), *The Internet of Things*, Springer-Verlag, Berlin, 2008.

[FRI 11] FRIEDMAN S.M., "Three Mile Island, Chernobyl, and Fukushima: An Analysis of Traditional and New Media Coverage of Nuclear Accidents and Radiation", *Bulletin of the Atomic Scientists*, vol. 67, no. 5, pp. 55–65, 2011.

[GER 99] GERSHENFELD N., *When Things Start to Think*, Henry Holt, New York, 1999.

[GOM 10] GOMEZ RODRIGUEZ M., LESKOVEC J., KRAUSE A., "Inferring Networks of Diffusion and Influence", *Proceedings of the 16th ACM SIGKDD International Conference on Knowledge Discovery and Data Mining*, Washington, United States, pp. 1019–1028, July 25–28, 2010.

[GOR 14] GORRE F., "Tremblement de terre, tsunami et accident nucléaire de la centrale de Fukushima: état des lieux des conséquences et des actions engagées trois ans après" , available at: http://www.ccr.fr/-/avis-expert-fukushima-3-ans-apres, 2014.

[GRU 04] GRUHL D., GUHA R., LIBEN-NOWELL D. *et al.*, "Information Diffusion Through Blogspace", *Proceedings of the 13th International Conference on World Wide Web*, New York, United States, May 17–22, 2004.

[HUG 09] HUGHES A.L., PALEN L., "Twitter adoption and Use in Mass Convergence and Emergency Events", *International Journal of Emergency Management*, vol. 6, no. 3, pp. 248–260, 2009.

[INT 05] INTERNATIONAL TELECOMMUNICATION UNION, The Internet of Things, ITU Report, 2005.

[KER 13] KERA D., ROD J., PETEROVA R., "Post-apocalyptic Citizenship and Humanitarian Hardware", in HINDMARSCH R. (ed.), *Nuclear Disaster at Fukushima Daiichi: Social, Political and Environmental Issues*, Routledge, London, 2013.

[KRA 12] KRANZ M., ROALTER L., MICHAHELLES F., "Things That Twitter: Social Networks and the Internet of Things", *What Can the Internet of Things Do for the Citizen (CIoT) Workshop at the 8th International Conference on Pervasive Computing (Pervasive 2010)*, Oldenburg, Germany, pp. 1–10, 2010.

[LER 10] LERMAN K., GHOSH R., "Information Contagion: An Empirical Study of the Spread of News on Digg and Twitter Social Networks", *Proceedings of the 4th Int'l AAAI Conference on Weblogs and Social Media (ICWSM 10)*, Washington, United States, pp. 90–97, 2010.

[LI 14] LI J., VISHWANATH A., RAO H.R., "Retweeting the Fukushima Nuclear Radiation Disaster", *Communications of the ACM*, vol. 57, no. 1, pp. 78–85, 2014.

[LIB 08] LIBEN-NOWELL D., KLEINBERG J., "Tracing Information Flow on a Global Scale Using Internet Chain-letter Data", *Proceedings of the National Academy of Sciences*, vol. 105, no. 12, pp. 4633–4638, 2008.

[LUN 13] LUNDEN I., "Mobile Twitter: 164M+ (75%) Access From Handheld Devices Monthly, 65% Of Ad Sales Come From Mobile", available at: http://social.techcrunch.com/2013/10/03/mobile-twitter-161m-access-from-handheld-devices-each-month-65-of-ad-revenues-coming-from-mobile/, 2013.

[ORE 05] O'REILLY T., "Web 2.0: Compact Definition?", available at: http://radar.oreilly.com/2005/10/web-20-compact-definition.html, 2005,

[PAL 09] PALEN L., VIEWEG S., LIU S.B. *et al.*, "Crisis in a Networked World Features of Computer-mediated Communication in the April 16, 2007, Virginia Tech Event", *Social Science Computer Review*, vol. 27, no. 4, pp. 467–480, 2009.

[PAL 10] PALEN L., ANDERSON K.M., MARK G. *et al.*, "A Vision for Technology-mediated Support for Public Participation & Assistance in Mass Emergencies & Disasters", *Proceedings of the 2010 ACM-BCS Visions of Computer Science Conference*, Edinburgh, United Kingdom, pp. 8:1–8:12, 2010.

[PET 11] PETERS I., KIPP M.E.I., HECK T. *et al.*, "Social Tagging & Folksonomies: Indexing, Retrieving… and Beyond?", *Proceedings of the American Society for Information Science and Technology*, vol. 48, no. 1, pp. 1–4, 2011.

[PLA 11] PLANTIN J.-C., "'The Map is the Debate': Radiation Webmapping and Public Involvement During the Fukushima Issue", *SSRN Electronic Journal*, September 2011.

[ROS 14] ROSE D., *Enchanted Objects. Design, Human Desire and the Internet of Things*, Scribner, New York, 2014.

[VAN 90] VAN DER PLIGT J., MIDDEN C., "Chernobyl: Four years later: Attitudes, risk Management and Communication", *Journal of Environmental Psychology*, vol. 10, pp. 91–99, 1990.

[VAN 07] VANDERFORD M., NASTOFF T., TELFER J. *et al.*, "Emergency Communication Challenges in Response to Hurricane Katrina: Lessons from the Centers for Disease Control and Prevention", *Journal of Applied Communication Research*, vol. 35, no. 1, pp. 9–25, 2007.

[VER 14] VERMESAN O., FRIESS P. (eds), *Internet of Things: from Research and Innovation to Market Deployment*, River Publishers, Aalborg, 2014.

[VIE 10] VIEWEG S., HUGHES A.L., STARBIRD K. *et al.*, "Microblogging During Two Natural Hazards Events: What Twitter May Contribute to Situational Awareness", *Proceedings of the 28th Conference on Human Factors in Computing Systems (CHI2010)*, Atlanta, United States, pp. 1079–1088, April 10–15, 2010.

[WHI 11] WHITE C., *Social Media, Crisis Communication, and Emergency Management: Leveraging Web 2.0 Technologies*, CRC Press, Boca Raton, 2011.

[XIA 12] XIA F., YANG L., WANG L. *et al.*, "Internet of Things", *International Journal of Communication System*, vol. 25, no. 9, pp. 1101–1102, 2012.

[ZAN 14] ZANELLA A., BUI N., CASTELLANI A. *et al.*, "Internet of Things for Smart Cities", *IEEE Internet of Things Journal*, vol. 1, pp. 22–32, 2014.

8

Connected Objects:
Transparency Back in Play

8.1. Introduction

The Internet of Things (IoT) is characterized by the spread of objects capable of automatically capturing and exchanging data in our environment. Rather than causing the appearance of new objects in our daily lives, the IoT represents the transformation of familiar objects with the goal of simplifying their operation and increasing the number of their functions. Extended to objects, the Internet modifies the way in which it is possible to contemplate our environment and perceive the elements that make it up. Promoters of the IoT consider each object a potential producer and consumer of data, and as an information appliance to accomplish a predefined and limited number of tasks. This extension of the Internet to objects belongs to the third age of computing history: ubiquitous computing is superseding the era of personal computers and that of mainframe computers.

The IoT modifies the status of the objects surrounding us by letting them adapt to different contexts and user profiles. As the father of ubiquitous computing, Mark Weiser, wrote: "If a computer merely knows what room it is in, it can adapt its behavior in significant ways without requiring even a hint of artificial intelligence" [WEI 99]. The IoT endows objects with the capacity to alter their own functioning. It expands the functions of everyday objects and gives them additional roles. Equipped with sensors, connected objects can

Chapter written by Florent DI BARTOLO.

control the deeds and actions of their users and exercise new monitoring and support functions[1].

Connected objects function based on low energy consumption that increases their autonomy. Autonomy is a property acquired progressively in relation to a certain context of use. It is built from raw data that is regularly collected, assembled and then analyzed on remote servers. The autonomy of connected objects lets their designers offer unprecedented services and imagine new experiences to live that do not require direct nor conscious interaction with a digital interface. It makes it possible to design objects that aspire to blend in with an environment by being totally transparent and imperceptible to the eyes of their users.

In this chapter, we propose to analyze the quest for transparency that seems to guide promoters of ubiquitous computing and of the IoT. The transparency of connected objects is expressed through the quasi-invisibility of their digital interfaces and data flows they are exchanging. It takes shape through the images that support their commercialization and presents them as artefacts in ethereal forms. The difficulty in perceiving these objects is also the consequence of the opacity that governs their operating mode: obfuscation has become a legitimate information management strategy that limits the attention it is possible to pay to their materiality.

8.2. Sensitive objects

The mode of existence of hypermedia objects has already been addressed in terms of opacity and transparency in the 1990s by Richard Grusin and Jay David Bolter, but the concept of opacity was not used to refer to a lack of clarity vis-à-vis the functioning of a technical object [BOL 00]. On the contrary, opacity was used to indicate a degree of visibility beyond which an interface becomes perceptible, graspable, by capturing the attention of its users. For these two researchers, hypermedia objects constantly oscillate between transparency and opacity: their erasure is constantly called into question, disrupted by their interfaces, which remind us of their existence through their elements of tabularity, but also accidental experiences.

---

1 For example, activity trackers such as "connected bracelets" offer to measure the intensity of our efforts or the quality of our sleep by considering different parameters such as the length of light, deep and REM sleep.

Jay David Bolter and Richard Grusin describe *transparent immediacy* as the quasi-invisibility which digital interfaces claim to have. Reading and navigation supports respond to the logic of immediacy that requires them to erase themselves to leave us alone in the presence of the thing being represented [BOL 00]. Accompanied by Diane Gromala, Jay David Bolter returns to this quest for transparency in a second work entitled *Windows and Mirrors* [BOL 03] that guides "experts in human-computer interactions" like Don Norman and Jakob Nielsen and for whom computers represent, according to the authors of this book, only information appliances.

The desire to make interfaces invisible has been clearly stated by the scientists working on the very first virtual reality devices [HOD 94]. Whatever their size and weight, virtual reality devices have the goal of turning us away from the technical object by promoting immersive experiences that anchor us to another reality. They are based on a feeling of presence that pushes us away from the world in which we live and temporarily integrates us into another. Current virtual reality peripheral devices (Oculus Rift, PlayStation VR, HTC Vive) carry the same promise. They recycle the imagery associated with the first virtual reality environments by inviting users to experience another world with the help of peripheral devices that have become less burdensome, but still don't completely and conclusively avoid "virtual reality sickness", a problem that commonly affects users of these devices and that is manifested in the form of headaches, nausea or even vomiting [LAV 00].

The peripheral devices of virtual reality are not the only interfaces presented by their designers as elements of an interactive device that are meant to disappear. This is also the case for connected objects, but the techniques used, like the goals, diverge. Connected objects become transparent by blending in with their users' everyday environment. Unlike virtual reality devices, ubiquitous computing is not trying to simulate a world, but to "improve", "enrich" and "enhance" the one in which we live using the help of machines spread throughout our environment and connected to each other [WEI 99]. This transparency is not the result of getting progressively accustomed to their omnipresence. It is the consequence of a unique mode of existence where it is not possible to perceive the presence of the devices clearly or to know their scope of action during contact or an interaction with their interfaces. The relative autonomy of connected objects also facilitates their integration into the environment and participates in their erasure. In that regard, we will attempt to call into question the oscillation between the transparency and opacity of digital interfaces as stated by Richard Grusin and

Jay David Bolter regarding peripheral devices and applications that largely escape our notice while their main function is to direct it.

Connected objects are meant to erase themselves in favor of experiences that require less direct interactions with their interfaces. They do not act like virtual reality devices, by immersion or by saturating their users' field of vision. They are not trying to make us forget our immediate environment. On the contrary, connected objects and the applications associated with them have been created to give us a more detailed representation of ourselves (the Quantified Self) and of our environment by capturing data directly from real life. This is the specific case of the application Google Fit, which is compatible with all Android Wear devices, and which makes it possible to "effortlessly track any activity. As you walk, run, or cycle throughout the day, your phone or Android Wear watch automatically logs them"[2]. It is enough to keep your phone with you for the data to be automatically collected and connected with different activities such as walking, running or biking. The user does not have to choose an activity or specify its duration. The connected object collects, with the help of a collection of sensors (accelerometer, gyroscope, microphone, GPS, barometer, etc.), data concerning the geographical location of its user and the movements that he is performing in the background, with the goal of distinguishing one athletic activity from another.

Connected objects have invaded many spaces (towns, hospitals, the home, clothing, automobiles) and activity areas such as health, home automation, fashion, art and military operations. Their uses are multiple, just like their operating methods: not all objects referred to as "connected" are connected to the Internet. They are not necessarily capable of exchanging data with most micro- or nano-computers present in their vicinity, since they don't use the same communication protocols. Each manufacturer of connected objects is currently trying to put its own protocols on the IoT market. Nevertheless, most of the main actors seem to share the same ambition: to design objects capable of simplifying our human-machine interactions (HMI) by demanding the least effort possible from their users. This involves removing interactions considered useless, as the 2015 edition of the Google I/O conference organized at the Moscone Center in San Francisco, California, has already shown[3]. At this event, several actors from

---

2 Google Fit - Fitness Tracking - Android Apps on Google Play. Internet source:
https://play.google.com/store/apps/details?id=com.google.android.apps.fitness&hl=en.
3 Google I/O 2015. Internet source: http://events.google.com/io2015/.

the Mountain View firm presented their medium-term (2020) vision of a connected world addressed to developers worldwide; a world in which each one of our activities is monitored with the goal of intensification via their personalization, our interactions with everyday objects such as a clock radio or a car: the time we wake up is automatically calculated according to our level of fatigue the night before and our use of time during the day while our means of transportation are capable of announcing their imminent arrival to us and offering us access to informative or entertaining content such as music tracks that are selected according to the tastes and sensibilities of the group we are part of in a given situation (car travel, meeting, family reunion)[4]. Connected objects are expected to make us more conscious of the presence of the people that surround us and the occurrence of events that we are likely to appreciate (and in which it is possible for us to participate) by giving us access to information adapted to our current situation, to a precise context, and without having made the demand previously.

Connected objects progressively acquire their autonomy. They adapt to the behavior of their users as well as their environment based on the data that they capture automatically, but also based on data that their users enter or correct. The application Google Fit, for example, makes it possible to modify the data captured by Android Wear devices to correct errors and enhance the quality of the information obtained[5]. Connected objects remain objects that it is possible to set up manually, but this setup does not have to be done each time they are used. Their user is instructed to evaluate their performance to improve their quality, and not activate or interrupt data capturing. Connected objects record by default their users' actions, like the Nest thermostat that automatically regulates the temperature of a house according to its inhabitants' way of life[6].

Designers are making the handling of digital objects more evident by taking advantage of the commercialization of new models for simplifying their interfaces. As John Maeda has demonstrated, this simplification is likely to take different shapes. It sometimes reverts to applying "thoughtful reduction" by progressively reducing the choices that are offered to the user

---

4 Google I/O 2015 – Making apps context aware: Opportunities, tools, lessons and the future. Internet source: http://www.youtube.com/watch?v=xgcj7VbDalk.

5 Google Fit –Google Fit support center. Internet source: http://support.google.com/fit/?hl=fr#6 223934.

6 Nest – This is the Nest thermostat. Internet source: http://nest.com/fr/thermostat/life-with-nest-thermostat/.

or even carrying out these choices for him [MAE 06]. This radical approach has contributed to the success of products such as the iPod Shuffle whose commercial launch in 2005 was accompanied by the slogan, "Random is the New Order"[7]. By deconstructing any order created during the creation of music playlists, the iPod invited its users to "lose control" and to "love it". Due to the loss of control that it creates, random access is as much a demonstration of the power that interfaces exert on how data is read, one example of the pleasure possible to get from a simplified mode of interaction: the random playing of audio files can induce a feeling of surprise in their listeners and produce pleasing arrangements. However, this kind of design cannot be chosen without greatly restricting users' freedom of action. Therefore, the functioning of connected objects is based, if only a little, on choices and adjustments made consciously by their owners.

Connected objects adapt to their environments. As part of a distributed architecture network, they carry out actions based on the analysis of raw data that they collect autonomously and collectively. Their manufacturer uses analytical models that make it possible to distinguish between different contexts of use in order to give birth to new forms of experiences. For example, the application Smart Lock[8] makes use of geographical data collected by several connected objects to identify "trusted places" and automatically unlock electronic devices such as tablets and smartphones when they enter the security perimeter of these places. The IoT allows applications to propose, based on data that has become more precise (such as localization data), new functions that automate operations carried out mechanically by their users. This done, the IoT contributes to reducing the attention that we pay to objects and their interfaces.

The declarations of designers like Bill Buxton (research director at Microsoft Research) demonstrate it: connected objects are intended to integrate perfectly into our environment to the point where we forget their existence completely[9]. Connected objects are seen by their designers as

---

7 Apple – iPod shuffle. Internet source: http://web.archive.org/web/20050112043302/ www. apple.com/ipodshuffle/.

8 Google Smart Lock. Internet source: http://get.google.com/smartlock/.

9 Bill Buxton: "In some sense, a successful interaction design would be transparent, almost invisible, to the point that the user would be almost unconscious of the experience until after it's over. It is just like magic. A good interaction design also needs to fit well within the society of appliances that surrounds it." Internet source: https://rslnmag.fr/cite/bill-buxton-the-best-interaction-design-is-transparent-almost-invisible/.

objects that should be made transparent to create the illusion of unmediated, direct contact between a piece of information and its recipient. A new object, a new application, must be capable of subtly taking its place in its user's environment in a way that frees their use from any hitches or frictions, to the point of sometimes giving a magical dimension to the experience of a technological device. Any trace of mediation must fade into the background.

8.3. The myth of transparency

Connected objects make it possible to diversify the forms in which a single piece of information is communicated to us. As dynamic data visualizations multiply the views that it is possible to have of the same phenomenon, connected objects participate to make any representation provisional, susceptible to being modified, altered, but also completed by new representations, available on additional supports. Connected objects exploit the power of the web's information systems and their data centers. Information does not have to appear in all its complexity. On the contrary, it can be reduced to a light signal, a variation in color, or a vibration as the artistic work of Julien Levesque shows,[10] exploring the poetic dimension of the IoT and the data flows to which it allows access, in keeping with the work carried out by artists like Natalie Jeremijenko at Xerox PARC in Palo Alto in the 1990s [WEI 96]. Connected objects diversify the forms of access to information all the while reinforcing its personalization: the information is adapted by default to a particular context as well as to a unique digital identity.

The IoT gives shape to a myth of transparency marked by the automation and computerization of data capture processes and the selection of information. The environment progressively becomes the interface, while the computational processes are relegated to the background and offer no visibility [KRA 07]. The myth of transparency has invaded spaces of communication and consumption. It is constructed by actors such as engineers, developers and interaction designers using expertise, but also intuitively. It is depicted in advertising and in movies. The myth of transparency takes shape through all of the images that present digital technologies in the form of technical objects quietly showing up in our environments, and are located by default on the edge of our gaze. The

10 Julien Levesque – Selected Works. Internet source: http://www.julienlevesque.net/.

imagery associated with the myth of transparency is perpetually being renewed. It evolves with the society whose future it envisages. Commercial videos created by companies like Microsoft and Apple give this myth a central place. They establish its contours as shown in the publicity film "Productivity Future Vision" produced in 2015 by Microsoft[11].

"Productivity Future Vision" offers a medium-term vision (between five and ten years) of the world of work: a world revolutionized by ubiquitous computing and invaded by screen surfaces. Screens are everywhere and nowhere at once. They no longer have a frame or thickness of their own: the surface of the smallest object present in a workplace or a home is susceptible to being used to receive information and communicate it to its surroundings. Beyond the objects that fill our living spaces, it is our dwelling places and workplaces themselves that fulfil the function of a screen and that are required to disappear, to become transparent, permeable to ubiquitous computing and its data flows. Each glass door, each wall is presented as a tactile surface ready to be activated, to respond to an imperative of connectivity. "Productivity Future Vision" gives us a vision of a world populated primarily by objects and not by human beings. In this world, where all information is always desirable and necessary, humans are nothing more than bit players or foils. Stripped of words, the actors have only the purpose of highlighting the fluid and frictionless HMIs on which the myth of transparency is based.

The myth of transparency has an influence on the design of the objects and services that we use daily. It is built around several themes that "Productivity Future Vision" highlights: collaborative work done remotely, "smart personal agents" (assistive intelligence), friction-free interfaces known as "natural interfaces" and the free circulation of data and people (fluid mobility). These themes have an impact on the design of user experiences. They keep the promise of better communication between human beings via the use of technologies capable of disappearing (to better bring their users together around a common activity), but also via the use of "proactive" objects designed to suggest options for their user and able to motivate behaviors:

> "Kat receives an invitation from Lola on her bracelet. Her personal agent proactively suggests options for her. She can

11 Microsoft – Productivity Future Vision. Internet source: https://www.microsoft.com/enterprise/productivityvision/default.aspx.

now use simple gestures to accept the invitation, rearrange her calendar, and book a space to prepare".

Thanks to the use of personal agents presented in the form of assistants, the act of creation becomes an activity devoid of friction (friction-free creativity): telling stories, organizing ideas or mining data happens effortlessly thanks to supports that automate these tasks and share their results. The circulation of data between connected objects is presented as being fluid and secure just like the adaptive environments in which their users evolve and which adapt to their presence by identifying them: "In the lab, the blackboard recognizes Kat's team as they enter the space. They can quickly 'rehydrate' the room with their project and resume where they left off". The myth of transparency has the effect of hiding the materiality of digital devices that cannot be clearly identified or distinguished from the environment in which they are placed. The use of the term *cloud computing* to evoke the exploitation of the processing power and storage belonging to server farms is an example of the ambiguity which surrounds the Internet architecture and the functioning of connected objects.

The functioning of connected objects is not apparent. Learning how to operate them involves gestures that their users are led to discover progressively by interacting regularly with them. Engineers and interaction designers use the term "natural interface" for digital interfaces that are not perceptible by default and remain that way while being used. This is particularly the case with the software interfaces of touch screens whose visibility is reduced to a minimum so that their users attribute functions that belong to an operating system to the support itself (a portable telephone, a tablet). The name "natural interface" is also assigned to software libraries that make it possible to create user interfaces that escape notice. For example, the NUI (Natural User Interface) software library provides access to data captured by Microsoft Kinect, a peripheral device which allows interaction with a computer using vocal commands and image and movement recognition[12]. In the field of digital arts, natural interfaces have been designed from the beginning of the 1990s by artists like Christa Sommerer and Laurent Mignonneau to develop interactive installations such as *Interactive Plant Growing* (1992) and *A-Volve* (1994). The concept of "natural interface" again goes back to HMIs that are carried out "naturally" which means without attention being paid to them, but it is also used to refer

---

12 Microsoft – Natural User Interface for Kinect for Windows. Internet source: http://msdn. microsoft.com/en-us/library/hh855352.aspx.

to the natural elements (plants, freshwater basins) that serve as physical interfaces and mitigate the presence of computer systems to which they are connected.

The transparency of digital interfaces is built on metaphors and design models that belong to the objects to which we are already accustomed. These imported models simplify HMIs by making them familiar, but they also have the effect of hiding the technological advance of ubiquitous computing by making it impossible to grasp the functioning of connected objects clearly or to propose forms of interaction that are likely to radically change our daily lives. The technological expansions related to the IoT become part of our use without us being able to consider their entire extent. The simplification of HMIs doesn't let us authenticate the advances of an era in the form of access and visibility that would be given to data flows or information. On the contrary, it places, without fanfare or ruptures, the new functions of connected objects in our daily lives to facilitate their adoption by new consumers.

Though they are attempting to enter our lives quietly, connected objects are disrupting our way of life, particularly regarding access to information. Connected objects multiply, through their sensibility to context and their automatization, the chances to access more varied and relevant information. Information no longer needs to be researched actively. It can be delivered in the form of notifications, such as alert messages that users of mobile applications receive. Push notifications represent one of the numerous mechanisms used by mobile platforms to inform their users of a planned event or remind them of it at the right time. Access to information is no longer necessarily the result of upstream research, but the consequence of actions carried out in the presence of connected objects and analyzed by private companies. In other words, connected objects give even more power to the world's largest Internet companies and importance to the calculations they perform to analyze the traces of our activities and respond to our interrogations[13]. Connected objects participate in the creation of reality. They organize and orient their users by providing access to information adapted to their identity and to the objects with which they are currently in contact

---

13 With the stated goal of improving the quality of its different services, Google is now asking its users the permission to record their search activity, the history of their positions, but also the history of information originating from devices to which they are connected (contacts, agendas, alerts, applications, music, films, books and other content) as well as their vocal and audio entries (to contribute to the recognition of their voices and the improvement of voice recognition). Internet source: http://myaccount.google.com/privacy.

(mobile telephones, tablets, connected watches). They have the capacity to broaden our horizons, but also, depending on the metrics which they follow, to put us in "filter bubbles" capable, conversely, of restricting our freedom.

The democratization of connected objects has been accompanied since the end of the 2000s by the publication of press articles announcing the end of the web in favor of new services considered more attractive and user-friendly [WOL 10]. The figure of the "flâneur" once used to describe a typical way of browsing information online no longer makes sense in the face of a multiplication of services providing almost instantaneous access to relevant information [MOR 12]. Although it is still possible to browse the web by following hyperlinks, the Internet has become principally an informational space that its users interrogate with the help of requests formulated automatically by applications and that are meant for programming interfaces (API). Connected to the databases of the web's information systems, programming interfaces return the results of our desired or involuntary HMIs directly to connected objects, so we don't have to browse the databases ourselves. The transparency of connected objects is based not only on their interconnection, but also on communication in real time that they can establish with information systems.

Despite the necessity for connectivity on which the IoT is based, there is still currently no "universal language" allowing any object to communicate easily with another, but companies like Google are promoting operating systems and communication protocols specially designed to facilitate exchanges between connected objects. In 2015, the Mountain View firm launched its operating system Brillo and invited software designers to contribute to the communication platform Weave dedicated to connected objects that accompany it. With the help of these new services, Google wants to create a planet-wide ecosystem of objects that use the same set of protocols for communicating with each other, and for interacting with their users and distant servers. The themes of the myth of transparency (immediacy of HMIs, interconnection, fluid circulation of data and objects' sensibility to context) are gathered together in the business discourse that accompanies the launch of these new services and carries the promise of a distributed intelligence[14].

---

14 Weave – Google Developers. Internet source: http://developers.google.com/weave/.

8.4. Transparency of interfaces and opacity of processes

Connected objects seek our attention to bring us information in forms that correspond to the image that a computer system has of us and our environment at a given moment. They operate in a "related environment" which makes their presence relevant[15] and reinforces the magical dimension of the interactions we have with digital devices. However, stories about the magical functioning of connected objects can also be explained by the lack of information that surrounds these functions and those of technologies that their presence in our environment engage such as radio-identification. Their users lack conceptual frameworks for understanding how these objects operate. They are no longer necessarily aware of the quantity and precision of the data that connected objects are capable of capturing, nor of the level of security and frequency of the communications that they have with servers belonging to private companies. As stated previously, connected objects do not just capture the data, they use the Internet to transmit it to distant machines on which it will be archived, then analyzed.

The intelligence of connected objects is intimately linked to this work of analyzing and processing data flows that can reveal on a largescale patterns, trends and unexpected correlations between different sociocultural phenomena. However, this work of data capture and behavior analysis cannot be envisaged without the consent and the control of the people to whom they are supposed to fully belong[16]. Automation of connected objects cannot be total without making their functioning incomprehensible and their use alienating. As Gilbert Simondon wrote in 1958, contrasting the figure of the automaton with that of the open machine, "the true perfection of machines, which we can say involves a high degree of technicality, does not correspond to an increase in automatism, but the opposite, since the functioning of a machine contains a certain margin of uncertainty. It is this margin that allows the machine to be sensitive to external information" [SIM 12]. The customization of connected objects is an opening factor, just

---

15 The effectiveness and relevance of the interactions that it is possible to have with connected objects are intimately linked to the capacity of our living spaces to adapt to their presence notably by taking into account their demand for connectivity.

16 The security of the data captured and transmitted by connected objects represents a true challenge to which Google is attempting to respond with the help of Weave: a communication platform that guarantees the encryption of data and that offers users the possibility of controlling the ways of accessing their information at a level described as "granular". Internet source: http://developers.google.com/weave/.

like their capacity to capture data originating in their environment or to take into account the presence of neighboring objects. This sensibility by which an object will resonate with its surrounding environment is for Gilbert Simondon at the root of all real technological progress.

However, the margin of uncertainty of connected objects is also strongly restricted by the opaque nature of their operating methods and the rules which they obey. The embodiment process by which the functioning of technical objects are evolving is held back by strategies set up to make them less accessible to their users' view, because this lack of visibility does not make it possible to imagine for these objects and their elements, a multifunctionality. Authenticating the innovation, identifying the potential of a technical device to imagine new uses for it, requires a certain degree of visibility. Connected objects need not only to be set up by their users, but also practiced and designed in order to extend their range and functionalities.

Making sense of ubiquitous computing to consider the technologies on which they are based, such as materials, requires investing time in practicing connected objects, as Timo Arnall's 2009 work around radio waves (to which RFID chips react) shows. For this designer, RFID technology has long remained misunderstood and disputed because of the invisible and involuntary nature of the interactions that it makes possible: "Once RFID antennas are hidden inside products or in environments, they can be invoked or initiated without explicit knowledge or permission"[17]. The invisibility of the radio waves represents a challenge for designers who must understand their properties (not least their form and their range) to design services that use RFID technology so it can have something new to offer in terms of HMIs, and to take care of their users by giving them the possibility of discovering the functioning of the objects with which they enter into contact.

The visibility that Timo Arnall gives to radio waves is sorely lacking today in the connected objects whose operations largely escape their owners. As Bruce Sterling has written, "the reader may be allowed to choose the casing of his smartphone and the brand of his vacuum cleaner, but the digital relation between the two of them is not his decision" [STE 14]. The user does not control the data exchanged between objects nor its commercial use. It is not the orchestra conductor that Gilbert Simondon describes, this

---

17 Timo Arnall – Immaterials: The Ghost in the Field. Internet source: http://www. nearfield.org/2009/10/immaterials-the-ghost-in-the-field.

"permanent organizer of a society of technical objects that require his presence" [SIM 12]. The user of a connected object integrates a network of relationships in which he is just one of many links. It implicitly participates in generating data whose commercial use and market value are unknown to him. Following the logic of "least-revealing means", upheld since the end of the 1990s by Lawrence Lessig concerning the personal data that private companies require we share with them to use their services [LES 00], today some businesses are trying to define new contract formats which would allow users of connected objects to reclaim their data. A good example is the company IF whose *Data Licences* (exhibited in 2016 at Somerset House, in London, during the exhibition *Big Bang Data*) let its users define not only the type of data that they agree to communicate with businesses from their connected objects, but also to fix the price and conditions of this sharing: "A data licence is a design pattern that puts people in control of their data, letting them set the rules of engagement. By answering a short series of straightforward questions, users customise their data licence to form a contract with the other party"[18].

The functioning of connected objects also eludes their users due to the proprietary software that they are attached to, and whose source code remains impenetrable. By forbidding access to their source code, proprietary software prevent users of connected objects from studying their functions and making modifications. Their users are beholden to the businesses that sell them and profit from the data that they capture. Connected objects work like black boxes. Their complexity is concealed behind interfaces that are user-friendly which prevent us from becoming aware of the networks into which they integrate us.

The opacity of the processes adds to the transparency of interfaces to create "enchanted objects" [ROS 15], which are objects whose functioning is intentionally hidden with the aim of surprising their users, and entertain them with the help of strategies that tend to create powerful illusions. However, the exploratory practices of designers like Timo Arnall show that the qualities of a connected object cannot be reduced to a magical aura that should surround its use in our environment, or to the obviousness and the simplicity of the gestures that its setup requires us to perform. Connected objects can also be seen as tools whose connectivity and use evolve in the hands of their users, within the scope of their own actions and desires. The act of considering connected objects as open objects whose connectivity

---

18 IF – Data licences. Internet source: https://projectsbyif.com/projects/data-licences.

remains to be defined requires us to be able as users to study the software that their functioning integrates.

The interfaces and the source code of proprietary software (in particular, graphics software) have become a real source of inspiration and reflection in the field of art. For example, *Software Art* exaggerates the automation of software and disrupts its functioning in a way that demonstrates their impacts on our lives as the artistic work of Adrian Ward or Adam Harvey shows [DIB 15]. The work of certain artists specifically *interrogates* the silent functioning and presence of connected objects in our environment. They design projects that renew the forms of reception and visibility given to the data that these objects collect, but also to the signals they receive and emit.

For example, in 2012, artist and engineer Julian Oliver created *The Transparency Grenade*. The connected object takes, as its name indicates, the shape of a transparent grenade that it is possible to activate in a public place in order to intercept data transmitted via wireless networks. The data captured is automatically sent to a private server that analyzes them to recover information (user names, IP addresses, fragments of emails, images, etc.) which make it possible to identify individuals[19]. Presented in June 2015 in Cergy for the *Data et moi* exhibition, the grenade was accompanied by a video projection exposing the data in the process of being exchanged by the exhibition's visitors to everyone via a Wi-Fi network called "PublicWireless". Julian Oliver does not use this object for his own purposes nor does he provide servers for others to use it. *The Transparency Grenade* is a critical research project that interrogates the volatility of the information we exchange by radio waves. The exploitation of the vulnerability of computer systems represents for this artist a form of denunciation that adds a performative dimension to the artistic gesture[20]. Exploiting the vulnerabilities of connected objects in order to give to their interfaces and the data that they exchange new forms of visibility is a path that many artists are taking today

---

19 Julian Oliver – The Transparency Grenade (2012). Internet source: http://transparency grenade.com/.

20 *The Critical Engineer considers the exploit to be the most desirable form of exposure.* (The Critical Engineering Working Group, *The Critical Engineering Manifesto*, 2011–2015). Internet source: http://criticalengineering.org/en.

and that finds an echo in the data sonification artworks made by Ryoji Ikeda[21], Nicolas Maigret[22] or Jean-François Blanquet[23].

The vulnerability of connected objects represents not only a great opportunity for the artists and designers who choose to open them up for new forms of experiences despite their status as black boxes. It also constitutes a serious threat to their promoters. The low security of connected objects barely guarantees the private nature of data that they seem so eager to capture, as the multiple vulnerabilities and hacking of the Thermostat Nest demonstrate.[24] A survey carried out by Capgemini Consulting and Sogeti High Tech in November 2014 of 100 businesses and start-ups involved in the development of products using the IoT demonstrates that connected objects are, by their designers' own admission, poorly secured[25]. The strength of passwords which make it possible to modify their parameters is low and the majority of them do not encrypt the data that they communicate to remote servers.

The IoT market will not be able to be developed without its main actors giving greater importance to the safeguarding of the objects and applications that they create, since their adoption by new communities of users is based in part on the trust that it is possible to have in them in terms of the capture and management of personal data. The world's largest Internet companies such as Alphabet have the ambition to keep the confidence of their users by establishing relationships that allow them to understand how the data that they agree to share is used, and the immediate benefits that they can expect:

> "From better commute options in Maps to quicker results in Search, the data we save with your account can make Google services a lot more useful to you. [...] Save your search activity on apps and in browsers to make searches faster and get

---

21 Ryoji Ikeda, *data.tron* [WUXGA version], 2007. Internet source: http://www.ryojiikeda.com/project/datamatics/.

22 Nicolas Maigret, *System Introspection*, 2002–2012. Internet source: http://peripheriques.free.fr/blog/index.php?/works/2010-system-introspection/.

23 Jean-François Blanquet, *Trafic de données*, 2015. Internet source: http://cromix.free.fr/.

24 Storm D., "Black Hat: Nest Thermostat Turned into a Smart Spy in 15 Seconds" *Computerworld*, 11/08/2014.

25 Capgemini Consulting – Sécurisation de l'Internet des Objets : la cybersécurité au cœur des objets connectés. Internet source: https://www.fr.capgemini.com/ressources/securisation-de-linternet-des-objets-la-cybersecurite-au-coeur-des-objets-connectes.

customised experiences in Search, Maps, Now and other Google products."[26]

The businesses of the IoT monetize the sharing of our data against the promise of a greater sensibility to context on the part of objects and services that they create. This increased state of connectivity translates into the personalization of user experiences, time saving, access to new functions and the automation of operations still accomplished mechanically. However, the sensibility of connected objects to their environment can also lead to new inconveniences, and prove to be extremely invasive as the *Haunted Machines* project introduced in February 2015 by Natalie Kane and Tobias Revell shows. For Natalie Kane, digital technologies suffer from their aptitude for personalizing our user experiences or rather for their inaptitude for fully considering the environment in which they evolve. They create anxiety by, for example, causing painful memories to resurface instead of the happy events that they are supposed to remind us of on social networks with the help of functions such as "On This Day"[27]. The technologies struggle to grasp the complexity of reality. The meaning of the multimedia data with which they put us in contact continues to escape them despite the use of powerful calculators responsible for giving meaning to our digital footprints: data mining algorithms can be used to bring up images of scenes of life previously shared on our screens, but they are not capable of predicting the reactions and feelings that their reappearance will cause, since these feelings depend on an infinity of variables that cannot be computed.

The artistic work of Lauren McCarthy highlights this incapacity of connected objects to fully take into account the environment in which they evolve and the emotions that their owners experience. The *pplkpr* project (pronounced *people keeper*) that she created in 2014 with Kyle McDonald explores the intrusive dimension of the IoT by offering an application to install on a mobile phone that was capable of making everyday decisions in our place based on the analysis of our heart rate recorded with the help of a "smart watch". With humor, the application *pplkpr* promises to optimize its users' social and professional relationships by automatically setting up meetings with people who arouse positive emotions, but also by limiting the visibility of individuals who cause feelings of anger or a rise in stress by

---

26 Google – Activity controls. Internet source: https://myaccount.google.com/activity controls.

27 Facebook Help Centre – On This Day. Internet source:https://www.facebook.com/help/439014052921484/.

ending the relationships that connect them on social networks[28]. This artistic project that takes the form of a mobile application interrogates the status of people who rely on wearable technology. Are they truly at the center of a process that aims to make them more conscious of their emotions, their environment and the people that surround them or are they in fact the opposite, instruments of data acquisition whose analysis and value only concern them indirectly? In order to answer this question, certain artists are addressing the power and wealth that the commodification of data collected by private companies generates and are designing new devices that allow them to gain control over the content they have chosen to put online. The artist Jennifer Lyn Morone, for example, created her own business with the stated goal of exploiting its market value. Presented in 2015 in Bâle, for the *Poetics and Politics of Data* exhibit, her work makes it possible to take stock of the incredible scope and finesse of the data that is collected on the scale of just one human being by Internet actors such as Google.

Connected objects modify the way we interact with our socio-technical environment. Their presence has an impact on our ways of being and communicating. They augment the exposure of our private lives and multiply the forms that the capture of our personal data take, this is why choosing their mode of existence should not be left to companies such as Amazon, Facebook, Alphabet, Microsoft or Apple. Connected objects do not necessarily have to escape our notice nor to enter silently in our everyday life. On the contrary, it is preferable to see them as objects whose visibility can be negotiated, and that are likely to produce frictions, to generate incidents. Connected objects cannot be considered "enchanted" or "benevolent" objects whose only purpose is to marvel us by reacting spontaneously to our presence or by paying attention to our activities. They need (in order to really serve their users) to be demystified, understood in all their complexity to evaluate their functioning, and possibly to redefine it starting from the introduction of new forms of HMI or connectivity.

Connected objects do not have to be seen as black boxes. On the contrary, the IoT can serve to open up our objects to original forms of connectivity that place their users at the center of new networks of relationships, and not on the edges, by considering them to be true orchestra conductors, or luthiers: capable not only of setting up connected objects, but also of creating or at least assembling them from elements such as the

---

28 Lauren McCarthy and Kyle McDonald, *pplkpr,* 2014. Internet source: http://lauren-mccarthy. com/pplkpr.

sensors and small single-board computers available on the market. To better serve the interests of their owners, connected objects must be connected with interfaces that make it possible to evaluate their behavior, but also to modify it by making it possible, for example, to specify the data that their users would like to record or the functions that they choose not to activate. The exploratory practices of Julian Oliver, Timo Arnall and Lauren McCarthy show that artists and designers have an active role to play in the definition of the IoT by not being content with using the connected objects created by the world's largest Internet companies, but by designing, on the contrary, their own tools.

8.5. Conclusion

The interfaces of connected objects can explicitly serve to define what is allowed in terms of interactivity by rigorously limiting the choices that it is possible for a user to make. However, their designers can also choose to underscore the functions that they judge more useful without necessarily deleting the others to guarantee the multifunctionality of their objects and keeping them from being considered simple utensils used to accomplish a specific task. When they are not used by their designers to require their users to accomplish a series of previously defined operations, digital interfaces have the capacity to invite their users to practice them the way it is possible to practice a musical instrument. They therefore encourage instrumental practice of digital technologies, that is to say an interminable, endless practice that is self-sustaining.

For the artist David Rokeby, the creator of numerous interactive installations, digital interfaces provide landscapes made up of hills, plains and mountains to travel through [ROK 98]. Their users explore hills and congregate on the plains, but they also sometimes engage in climbing of mountains. They therefore need help to progress in their ascension. A connected object cannot in this regard have transparency as a goal. An interface must be visible to its users to give them control over their upward journey. The interface of a connected object does not, however, have to be constantly deployed. It can on the contrary adapt to the needs of the user and appears in forms that take into account the nature of the support on which it will be made perceptible. Connected objects must become graspable through their interfaces when their users deem appropriate, like devices that represent for Mark Weiser and John Seely Brown "calm technologies" [WEI 96].

This state of visibility is not based on any technology and represents one of the main principles of interaction design laid out by Don Norman [NOR 10]. It goes against the myth of transparency, since it requires digital interfaces to appear as what they are: technical objects that provide access to multimedia data and whose functioning must be able to be comprehended to be progressively mastered. The transparency of connected objects is also to be called into question, since it is built on a form of control established and exercised by private actors. The smallest word, the least action, can be detected and become the object of computer processing with the goal of simplifying HMIs and not relying anymore on individuals to find data adapted to a user's profile or a particular context. The myth of transparency is a fiction in which a pervasive state of surveillance is responsible for the fluidity and ease with which we interact with digital devices, sometimes without even being aware of it.

Even if the implicit nature of these interactions sometimes provides real benefits in terms of user experience, it cannot in any case become the norm. In a time when our personal data is being massively indexed and analyzed, as much by government agencies as by private businesses, it is imperative that we can distinguish the digital devices from the environments in which they are immersed and that we can understand their interfaces in order to control access to our data and to know how it is being exploited. No technological device should be designed to escape our notice, but on the contrary, should be capable of being observed, at the request of its users, in all its complexity (even if this state does not correspond to its initial mode of presentation). Like an origami figure, a digital interface needs to be able, in the hands of its users, to be endlessly unfolded and refolded.

8.6. Bibliography

[BOL 03] BOLTER J.D., GROMALA D., *Windows and Mirrors: Interaction Design, Digital Art and the Myth of Transparency*, MIT Press, Cambridge, 2003.

[BOL 00] BOLTER J.D., GRUSIN R.A., *Remediation: Understanding New Media*, MIT Press, Cambridge, 2000.

[DIB 15] DI BARTOLO F., "Déjouer les interfaces" *Interfaces numériques*, vol. 4, no. 1, pp. 57–70, February 2015.

[GRA 09] GRAU O., "Living Habitats – Immersive Strategies", in SOMMERER C., MIGNONNEAU L. (eds), *Interactive Art Research*, pp. 170–175, Springer, New York, 2009.

[HOD 94] HODGES L.F., ROTHBAUM B.O., KOOPER R. *et al.*, Presence as The Defining Factor in a VR Application: Virtual Reality Graded Exposure in the Treatment of Acrophobia, GVU Technical Report, Georgia Institute of Technology, Atlanta, 1994.

[HUY 14] HUYGHE P.-D., *A quoi tient le design?*, De l'incidence Editeur, Grenoble, 2014.

[KRA 07] VAN KRANENBURG R. *The Internet of Things. A Critique of Ambient Technology and the All-seeing Network of RFID*, Institute of Network Cultures, Amsterdam, 2007.

[LAV 00] LAVIOLA J.J. JR., "A Discussion of Cybersickness in Virtual Environments", *SIGCHI Bulletin*, vol. 32, no. 1, pp. 47–56, January 2000.

[LES 00] LESSIG L., *Code and Other Laws of Cyberspace*, Basic Books, New York, 2000.

[MAE 06] MAEDA J., *The Laws of Simplicity: Design, Technology, Business, Life*, MIT Press, Cambridge, 2006.

[MOR 12] MOROZOV E., "The Death of the Cyberflâneur", *The New York Times*, February 4, 2012.

[NOR 10] NORMAN D.A., NIELSEN J., "Gestural Interfaces: A Step Backward in Usability", *Interactions*, vol. 17, no. 5, pp. 46–49, September 2010.

[NOR 98] NORMAN D.A., *The Invisible Computer*, MIT Press, Cambridge, 1998.

[ROK 98] ROKEBY D., *The Construction of Experience: Interface as Content*, Addison-Wesley, New York, 1998.

[ROS 15] ROSE D., *Enchanted Objects: Innovation, Design, and the Future of Technology*, Scribner, New York, 2015.

[SIM 12] SIMONDON G., *Du mode d'existence des objects techniques*, Aubier, Paris, 2012.

[STE 14] STERLING B., *The Epic Struggle of the Internet of Things*, Strelka Press, Moscow, 2014.

[WEI 99] WEISER M., "The Computer for the 21st Century", *ACM SIGMOBILE Mobile Computing and Communications Review*, vol. 3, no. 3, pp. 3–11, 1999.

[WEI 96] WEISER M., BROWN J.S., "Designing Calm Technology", *PowerGrid Journal*, vol. 1, 1996.

[WOL 10] WOLFF M., ANDERSON C., "The Web Is Dead. Long Live the Internet", *WIRED*, August 17, 2010.

9

Status of the Body within the Internet of Things: Revolution or Evolution?

9.1. Introduction

The Internet of Things (IoT) concerns the control of the physical world, made possible in numerous areas of activity by a processing chain that starts with a material object that measures its environment and returns the information to another object or to a central data integration system that is also capable of Big Data analyses. Effective control over the physical world is done either through the recovery of consolidated information, in a form that is easy for a human or a machine to interpret, or by feedback on objects equipped with activators. The Internet of Things therefore "modifies" "the way in which it is possible to perceive our environment and to interact with the objects that populate it" [DIB 15, p. 76].

Beyond the strictly industrial world, the IoT applies to the domain of wearables, home automation tools or hand-held objects, in the field of the individual and his/her environment, but also the social organization of a country. As a result, it poses a large number of corresponding questions, such as the control over behavior, the question of social acceptability and presence in the physical world and presence in the user's real and perceptive body [WEI 99].

In this article we explore the status of the body within the Internet of Things. We begin by studying the body in the field of sports and e-health.

Chapter written by Evelyne LOMBARDO and Christophe GUION.

9.2. Presence and absence of the body in the field of sports and e-health

The IoT creates the individual's image and modifies the individual's behavior in return. What connected objects show me about myself is superimposed over traditional images. I am offered an enhanced representation of previously unknown physical and physiological characteristics: I was this face and this body, and now I am also aware of being this heart that beats at a certain rate during physical activity, this mass that evolves according to this body weight curve, this breath that possibly puts me outside the norm of a certain criterion, while I control my sleep to be in good shape the next day. I monitor my characteristics and study my scalable curves. What was missing, inside my body but outside of my mind's reach, becomes suddenly visible, readable on a screen and absorbs my entire mind by its presence. The possibility that I have self-control over my own bodily, physical or medical evolution makes my body more present to me, but this perceptive body that is transformed into curves on a screen, is not an active, real body.

As long as everything is going well, "wellness" and m-health encourage the individual to engage with himself in a narcissistic way. In the field of sports, the IoT measures performance and encourages the individual to surpass him/herself. With the fulcrum of social networks, the quest for performance is enriched through comparison with others' results. As the individual ages, it is no longer a question of noting progress but of limiting regression; tools for comparison won't hesitate to compliment relative athletic results (such as, "for your age, that's great").

When the body fails, indicators emitted from objects will more likely go toward e-health devices, that is medical databases, which make the body traceable.

9.3. The traceability of the body or the integration of data by a digital coach

The IoT potentially allows me to track my experiences, my behaviors and my attitude. This is merely a question of sensors and analysis; my scalable identity card provided to me by my production of data: "the number of times, the date, the place where..." I sneeze; I blow my nose; I yawn; I smile; I laugh; I cry; I talk; I am in motion; I am immobile; I interact; I write; I read; I drive; I am with friends; with family and at the office.

Environmental characteristics can also be recorded simultaneously: noise level; air quality; humidity; pollen or light. Physiological characteristics are all also traceable: heart rate; perspiration; shortness of breath; speed of movement or sleep. The example of Chris Dancy, the hyper-connected human who monitors himself day and night through the use of connected objects is an extreme example of this Individu-Data [MER 13].

Thus, the algorithm, fed over days and years by data from connected objects, analyzes my behavior and my physiology within my relationship with the environment.

The IoT provides the algorithm with information, and the algorithm gets to know me, either because I provide it with my preferences or because it patiently analyzes my behavior; it becomes capable of providing value in the form of recommendations compatible with my previous tastes or reactions.

Beyond the individual person, if an entire population is connected, all of the ingredients are present to work on a "Big Data" analysis of this population's IoT data in order to deduce characteristics and patterns. The individual can then, using a digital coach that can take the form of a tablet application, benefit from comparing his/her behavior to this population's.

This raises the question of norms, and therefore the treatment of anything outside these norms and the social pressure that they cause, as well as of individual behavior induced by the knowledge of my situation in comparison to myself at another time, or my situation in relation to other populations.

The question of the individual relationship with this digital coach and the acceptable conditions for control is also raised. Systems permitting the geolocalization of the body could thus make up an "Everyware," as Adam Greenfield [GRE 14] calls it, a place where the body is always entirely traceable without the right to be forgotten.

9.4. The IoT creates a flow of information around the body: a present, readable and traceable cluster

The IoT involves objects that emit or receive a flow of information. If we design bursts of information in relation to the individual and coming from different objects that measure him/her both as a body and as an active being, we then make observations within the time and space which we propose to define by the term "Cloud" (body or thing-related data cloud).

We define the Cloud as the collection of the flows of information about a user issuing from or going toward each emitting or receiving sensor/actuator, submitted – voluntarily or involuntarily – to be measured by the sensor.

The information regularly transmitted concerns physical parameters (geolocalization, podometry, still and animated images); behaviors and elements of social life (sleep, facial expressions, eyes, speaking time, body dynamics, management of interlocutors), physiological elements (body temperature, heart rate, salt levels, glucose, amino acids of emunctories). At this stage we should note that a new category of sensors could see rapid growth depending on the progress of plastic surgery that introduces inorganic elements in place of biological organs: artificial hearts, hands, arms, eyes, etc.

The Cloud is therefore the flow of information relative to an individual or to another object, such as for example a home, observed through the IoT. The Cloud is made up of outgoing messages from sensors and incoming messages to actuators. The sensors and actuators in question are either worn or used by the individual in a self-measuring way (wearables and m-health tools), or within the individual's reach by voluntary peering (in the IoT of home automation or in a car).

Depending on the IoT system, the messages could be emitted voluntarily (sending orders, confirmations or NFC identifications), or involuntarily (geo-fencing which takes into account a geographical border crossing to launch an action, send an alert, etc.). The Cloud also includes information captured from certain Smart Building or SmartCity sensors such as occupancy sensors or cameras without voluntary retrieval on the part of the individual – and at this time without automatic recognition.

Finally, the Cloud has a strong relationship with a group of information systems whose role is to provide a service to the individual whether by acting in response (opening a door, turning on a fan, suggesting mood music, triggering an alert, etc.) or in displaying information on an interface (such as a smartphone screen). Its information systems are the different Clouds (storage, calculation, analyses) which each manage a specialized service.

The weak operational point that makes value creation difficult for the IoT paradoxically resides in its inventiveness and the multiplication of specialized services, which are provided and managed separately by diverse editors, each one using its own technologies and each one taking care to remain in charge of its own data flow. How is it possible to transform these overlapping partial flows and construct a complete, coherent and clustered view of the individual?

9.5. The body in interaction: sharing Clouds to inform the informational environment

A growing portion of our professional activities, along with a significant part our personal lives, involves remote interactions, which provide limited communication compared to face-to-face communication. The tools countering this degradation of quality are insufficient.

Even before attempting to begin a conversation, I can rely on Microsoft Communicator, which, verifying in real time what is on my future interlocutor's personal calendar in the Outlook messaging system, sends me a green signal indicating that my interlocutor is "available". I can then try to communicate via the telephone, with the only indication being simply the fact that he is not in a scheduled meeting or on the telephone. Currently, the form of remote communication that gets closest to face-to-face communication is videophony. Aside from the fact that it is not yet widespread, it is not always practical due to the asymmetry of the interlocutor's situation. For example, in a car, on public transportation, in a place with no privacy, etc.

To improve communication conditions by imitating the richness of face-to-face communication, interchanging Clouds is conceivable. The knowledge of the interlocutor's Cloud in real-time makes it possible to put the elements in context. The person who communicates via telephone from his/her car would provide his/her interlocutor – with the help of an interface that remains to be defined – with elements of their driving situation (the ideal would be to provide real-time footage of the driving environment), for example the elements of relative speed, proximity of vehicles, speed of gyration and acceleration. The interlocutor would then be able to better demonstrate empathy by understanding the contextual elements that explain why communication is not as smooth as during a call made from the comfort of the office. Chris Dancy's "Innernet," that place where the individual interacts totally with the environment of connected objects, listening to their feedback on our experiences does not seem far, and "in this vision of interaction design, the body and the environment become interfaces. And identity (of a person) will be defined by this interaction" [CAT 15, p. 95].

9.6. Clouds, persistence and trust: a mapped body without the right to be forgotten

A police investigation involves finding physical evidence: transit points, actions or exchanges of information. In the era of the IoT, sensors send a cluster of information permanently to the Clouds, archiving centers for the state of the physical world; its evolution. It also concerns humans: physical and physiological characteristics as well as those of position and movement. Potentially, in the time of Big Data, it becomes possible, and is those of sometimes required, to make use of the reconstruction of a consolidated vision from a precise extraction isolating the individual Cloud.

The innovation lies in the precision within time and space, coupled with the potentially very long-term persistence of data recorded in the Cloud. This convergence is enriched even more by the capacity for isolation and cross-referencing data which serve Big Data analyses.

Will it become more difficult to change your mind? Will we reflect/think more before acting? How will we control any deviation? Will behavioral norms change?

The trust that the user gives to digital companies still has to be established. Today, each one describes itself as a trusted third party, but it falls to businesses to become trustworthy actors by working on demonstrations, and behaviors vis-à-vis users. What degree of control over data should the user have and what do the laws or regulations say about it? What digital services should be developed to reassure the user?

Imagine a service that guarantees encryption of all information that comes from a user's Cloud. The encryption key could be, for example, linked to the user's smartphone: all of the information IoT emits near the smartphone would be subject to the key that allows de-encryption of the IoT cluster (including the camera). During an *a posteriori* analysis, and at a specific request (rogatory letters or for particular actors), the information could be de-encrypted by these actors only by voluntary action by the person whom the Cloud involves.

However, it should be stated that at this point encryption is almost impossible given the modest level of security implemented in the IoT. The information encryption systems that come from the strongest element (the smartphone) are themselves at the mercy of hackers, state intelligence services and security start-ups dedicated to decipher the keys. Will the individual still be able to control the data that he transmits, whether voluntarily or involuntarily?

9.7. The body, an object communicating between hyper-control and non-control

Connected objects created to communicate are primarily wearables (for example, a step-counting bracelet); they were born with the IoT.

However, not every connected object was created first and foremost to communicate: another class of communicating objects is becoming more popular, the result of adding sensors and communicating modules to objects whose primary function is not to communicate (a soda machine that sends out an alert when it reaches the threshold where it needs to be refilled).

Talking about the Internet of things (IoT) also means picturing the attachment of objects to the Internet through access networks and a gateway. It also means talking about data that comes from objects that supply the service platform. In the end, this data is processed by algorithms that provide value-added services, which is the purpose of the IoT for businesses: restitution of consolidated information, feedback to an automaton or an actuator, etc.

The hyper-connected human [CAT 15] produces data. Actimetric sensors measure his state of mobility, speed of movement and the time spent on these activities. Cardiac activity is measured by a cardio-frequency meter. The body becomes hyper-controlled and we enter progressively into the culture of the "Quantified-Self," a slow and insidious practice of measuring of the body that is spread through networks [LAM 14].

The effects of the IoT have therefore reached the private sphere, with the SmartHome – the world of wellbeing, energy optimization, and with the increasing power of portable connected instruments that, in the area of sports, of complete self-control (nutrition, activities, sleep, weight, cardio, breathing, etc.) influence the individual behavior of the enhanced human. The IoT facilitates the medical transition being imposed on our societies by promising us that we can remain at home and independent longer as seniors. But at the same time this auto-control of the body is mirrored by a lack of control over data transferred by the user about himself. In this way, the IoT, "lets you choose your smartphone case and your brand of vacuum cleaner, but not the relationship that links them to each other" [STE 14], the paradox of a body that is controlling and controlled at the same time.

9.8. Conclusion

At the end of this article we can once again ask the question: is the IoT an evolution or a revolution?

The IoT is the art of capturing data measurements, and transmitting these measurements remotely. The original goal of the IoT was to analyze, better understand and finally adopt, in the organic, physical or digital world, the correct reaction. An information-analysis-reaction loop, classic in the industrial world and that constitutes a technological evolution.

The effects of the IoT are therefore:

– to optimize the functioning of human processes at every scale;

– to increase the working weight within the industrial activity itself;

– to affect the entire action-reaction problem at the root of the behavior of a mechanical-software system, an individual or a social entity;

– to encourage the personalization of the user's treatment for every service based on individual behavior (risk management);

– at the behavioral level, to encourage normative effects in the individual and collective spheres.

Nevertheless, the IoT also constitutes revolution in the public and private spheres. If the connected car, which no longer requires a human driver, becomes a living space, certain elements in the arrangement of public transportation and workplaces must be modified.

The IoT transforms the representation of the physical world, and it adds unexplored dimensions to the world via, for example, data-visualization; one good approach involves the cartography made possible by the diffuse presence of billions of telephones. Smartphones leave traces of their state and of their routes when they travel with their owners. In this way the geolocalization of portable telephones makes it possible to create heat maps of the density of people carrying telephones over a period of time [GUI 15]. Cross-referencing these maps with precise events makes it possible to study group behavior in response to the event. On the individual level, analyzing the sequence of the "spectator's" actions [WEI 99] in front of his screen, coupled with the analysis of his attention by means of objects (a camera) will also allow behavioral study within the information-reaction loop. These cartographies involving telephones and smartphones herald other maps, involving status data from biological, medical, environmental, etc. sensors.

We have thus entered into the era of ubiquitous computing [WEI 93] where "the deepest technologies are those which have become invisible. Technologies that, tied together, form the fabric of our daily lives to the point of becoming inseparable" [WEI 91, p. 94].

In this context, questions about the body and human otherness submit to current tastes: if the human and the body are transformed into data, how will humans still be confronted with the other? And if the human is no longer confronted with otherness, will it still remain human? [REN 14].

9.9. Bibliography

[CAT 15] CATOIR-BRISON M.-J., "Quand le corps de Chris Dancy devient un objet connecté spectaculaire", *Actes du colloque H2PTM'15,* ISTE Editions, London, 2015.

[DIB 15] DI BARTOLO F., "Transparence et opacité des objects communicants", *Actes du colloque H2PTM'15*, ISTE Editions, London, 2015.

[GRE 14] GREENFIELD A., "All watched over by machines of loving grace: Some ethical guidelines for user experience in ubiquitous-computing settings", *Boxes and Arrows*, 2014.

[GUI 15] GUION C., LOMBARDO E., "Analyse de la cartographie par la géolocalization à l'heure de l'Internet des Objets", *Actes du colloque H2PTM'15*, ISTE Editions, London, 2015.

[LAM 14] LAMONTAGNE D., "La culture du moi quantifié-le corps comme source de données", *ThotCursus*, available at: http://cursus.edu/article/22099/ culture-moi-quantifie-corps-comme-source/#.Vv_ tgHocNT8, May 26, 2014.

[MER 13] MERZEAU L., "L'intelligence des traces", *Intellectica*, no. 59, pp. 115–135, 2013.

[REN 14] RENUCCI F., LE BLANC B., LEPASTIER S. (eds), "L'autre n'est pas une donnée, Altérités, corps et artefacts", *Hermès La revue*, no. 68, 2014.

[STE 14] STERLING B., *The Epic Struggle of the Internet of Things*, Strelka Press, Moscow, 2014.

[WEI 91] WEISER M., "The Computer for the XXIe Century", *Scientific American*, vol. 265, no. 3, pp. 3–11, 1991.

[WEI 93] WEISER M., "Hot Topics: Ubiquitous Computing", *IEEE Computer*, vol. 26, no. 10, pp. 71–72, 1993.

[WEI 99] WEISSBERG J.L., *Présence à distance. Déplacements virtuels et réseaux numériques: pourquoi nous ne croyons plus à la télévision*, L'Harmattan, Paris, 1999.

List of Authors

Aymeric BOUCHEREAU
ELLIADD
University of Franche-Comté
Besançon
France

Nasreddine BOUHAÏ
Paragraphe
Univeristy of Paris 8
Saint-Denis
France

Florent CARLIER
CREN
Université du Maine
Le Mans
France

Marie-Julie CATOIR-BRISSON
Institut ACTE UMR 8218
University of Nîmes
Nîmes
France

Carole CLOSEL
CORHIS
Paul-Valéry University
Montpellier
France

Marta DENISCZWICZ
Department Of Information Science
Federal University of Santa Catarina
Florianópolis
Brazil

Florent DI BARTOLO
LISAA
University of Paris-Est
Marne-la-Vallée
France

Audilio GONZALES-AGUILAR
IRSIC
Aix Marseille University
Marseille
France

Christophe GUION
Orange
Marseille
France

Moisés LIMA DUTRA
Department of Information Science
Federal University of Santa Catarina
Florianópolis
Brazil

Evelyne LOMBARDO
LSIS, UMR 7296, CNRS
Kedge Business School
Toulon
France

Adilson Luiz PINTO
Department of Information Science
Federal University of Santa Catarina
Florianópolis
Brazil

Valérie RENAULT
CREN
Université du Maine
Le Mans
France

Alexandre RIBAS SEMELER
Department of Information Science
Federal University of Santa Catarina
Florianópolis
Brazil

Ioan ROXIN
ELLIADD
University of Franche-Comté
Besançon
France

Imad SALEH
Paragraphe
University of Paris 8
Saint-Denis
France

Antonin SEGAULT
ELLIADD
University of Franche-Comté
Besançon
France

Federico TAJARIOL
ELLIADD
University of Franche-Comté
Besançon
France

Index

Other titles from

in

Information Systems, Web and Pervasive Computing

2017

DUONG Véronique
Baidu SEO: Challenges and Intricacies of Marketing in China

LESAS Anne-Marie, MIRANDA Serge
The Art and Science of NFC Programming
(Intellectual Technologies Set – Volume 3)

REYES-GARCIA Everardo, BOUHAÏ Nasreddine
Designing Interactive Hypermedia Systems
(Digital Tools and Uses Set – Volume 2)

SAÏD Karim
Asymmetric Alliances and Information Systems:Issues and Prospects
(Advances in Information Systems Set – Volume 7)

SZONIECKY Samuel, BOUHAÏ Nasreddine
Collective Intelligence and Digital Archives: Towards Knowledge
Ecosystems
(Digital Tools and Uses Set – Volume 1)

2016

BEN CHOUIKHA Mona
Organizational Design for Knowledge Management

BERTOLO David
Interactions on Digital Tablets in the Context of 3D Geometry Learning
(Human-Machine Interaction Set – Volume 2)

BOUVARD Patricia, SUZANNE Hervé
Collective Intelligence Development in Business

DAUPHINÉ André
Geographical Models in Mathematica

EL FALLAH SEGHROUCHNI Amal, ISHIKAWA Fuyuki, HÉRAULT Laurent,
TOKUDA Hideyuki
Enablers for Smart Cities

FABRE Renaud, in collaboration with MESSERSCHMIDT-MARIET Quentin,
HOLVOET Margot
New Challenges for Knowledge

GAUDIELLO Ilaria, ZIBETTI Elisabetta
Learning Robotics, with Robotics, by Robotics
(Human-Machine Interaction Set – Volume 3)

HENROTIN Joseph
The Art of War in the Network Age
(Intellectual Technologies Set – Volume 1)

KITAJIMA Munéo
Memory and Action Selection in Human–Machine Interaction
(Human–Machine Interaction Set – Volume 1)

LAGRAÑA Fernando
E-mail and Behavioral Changes: Uses and Misuses of Electronic Communications

LEIGNEL Jean-Louis, UNGARO Thierry, STAAR Adrien
Digital Transformation
(Advances in Information Systems Set – Volume 6)

NOYER Jean-Max
Transformation of Collective Intelligences
(Intellectual Technologies Set – Volume 2)

VENTRE Daniel
Information Warfare – 2ⁿᵈ edition

VITALIS André
The Uncertain Digital Revolution

2015

ARDUIN Pierre-Emmanuel, GRUNDSTEIN Michel, ROSENTHAL-SABROUX Camille
Information and Knowledge System
(Advances in Information Systems Set – Volume 2)

BÉRANGER Jérôme
Medical Information Systems Ethics

BRONNER Gérald
Belief and Misbelief Asymmetry on the Internet

IAFRATE Fernando
From Big Data to Smart Data
(Advances in Information Systems Set – Volume 1)

KRICHEN Saoussen, BEN JOUIDA Sihem
Supply Chain Management and its Applications in Computer Science

2013

BERNIK Igor
Cybercrime and Cyberwarfare

CAPET Philippe, DELAVALLADE Thomas
Information Evaluation

LEBRATY Jean-Fabrice, LOBRE-LEBRATY Katia
Crowdsourcing: One Step Beyond

SALLABERRY Christian
Geographical Information Retrieval in Textual Corpora

2012

BUCHER Bénédicte, LE BER Florence
Innovative Software Development in GIS

GAUSSIER Eric, YVON François
Textual Information Access

STOCKINGER Peter
Audiovisual Archives: Digital Text and Discourse Analysis

VENTRE Daniel
Cyber Conflict

2011

BANOS Arnaud, THÉVENIN Thomas
Geographical Information and Urban Transport Systems

DAUPHINÉ André
Fractal Geography

2009

BONNET Pierre, DETAVERNIER Jean-Michel, VAUQUIER Dominique
Sustainable IT Architecture: the Progressive Way of Overhauling Information Systems with SOA

PAPY Fabrice
Information Science

RIVARD François, ABOU HARB Georges, MERET Philippe
The Transverse Information System

ROCHE Stéphane, CARON Claude
Organizational Facets of GIS

2008

BRUGNOT Gérard
Spatial Management of Risks

FINKE Gerd
Operations Research and Networks

GUERMOND Yves
Modeling Process in Geography

KANEVSKI Michael
Advanced Mapping of Environmental Data

MANOUVRIER Bernard, LAURENT Ménard
Application Integration: EAI, B2B, BPM and SOA

PAPY Fabrice
Digital Libraries

2007

DOBESCH Hartwig, DUMOLARD Pierre, DYRAS Izabela
Spatial Interpolation for Climate Data

SANDERS Lena
Models in Spatial Analysis

2006

CLIQUET Gérard
Geomarketing

CORNIOU Jean-Pierre
Looking Back and Going Forward in IT

DEVILLERS Rodolphe, JEANSOULIN Robert
Fundamentals of Spatial Data Quality

Printed and bound by CPI Group (UK) Ltd, Croydon, CR0 4YY

27/10/2024

14580730-0001